The Human Services Internship
Getting the Most from Your Experience

Second Edition

Pamela Myers Kiser

Elon University

THOMSON

BROOKS/COLE

Australia • Brazil • Canada • Mexico • Singapore • Spain
United Kingdom • United States

THOMSON

BROOKS/COLE

The Human Services Internship: Getting the Most from Your Experience, Second Edition
Pamela Myers Kiser

Senior Acquisitions Editor: Marquita Flemming
Assistant Editor: Samantha Shook
Editorial Assistant: Meaghan Banks
Technology Project Manager: Julie Aguilar
Marketing Manager: Meghan McCullough
Marketing Communications Manager:
   Shemika Britt
Project Manager, Editorial Production:
   Christy Krueger
Creative Director: Rob Hugel
Art Director: Vernon Boes

Print Buyer: Linda Hsu
Permissions Editor: Roberta Broyer
Production Service: Aaron Downey/
   Matrix Productions Inc.
Copy Editor: Ann Whetstone
Cover Designer: Brenda Duke
Cover Printer: Thomson West
Compositor: International Typesetting
   and Composition
Printer: Thomson West

Library of Congress Control Number:
2006930936

ISBN-13: 978-0-495-09226-1
ISBN-10: 0-495-09226-6

Thomson Higher Education
10 Davis Drive
Belmont, CA 94002-3098
USA

For more information about our products,
contact us at:
**Thomson Learning
Academic Resource Center
1-800-423-0563**
For permission to use material from this
text or product, submit a request online at
http://www.thomsonrights.com.
Any additional questions about
permissions can be submitted by
e-mail to **thomsonrights@thomson.com.**

To my family

Contents

Chapter 2
GETTING STARTED

21

Chapter 3
GETTING ACQUAINTED

41

Chapter 4
LEARNING TO LEARN FROM EXPERIENCE: THE INTEGRATIVE PROCESSING MODEL

69

Chapter 5
USING SUPERVISION   93

Chapter 6
COMMUNICATING WITH CLIENTS   109

Chapter 7
DEALING WITH DIVERSITY
141

Chapter 8
DEVELOPING ETHICAL COMPETENCE
169

Chapter 9
WRITING AND REPORTING WITHIN
YOUR FIELD AGENCY   191

Chapter 10
MANAGING YOUR FEELINGS AND
YOUR STRESS   219

Contents

Chapter 11
TROUBLE-SHOOTING   243

Chapter 12
ENDING YOUR INTERNSHIP   271

Chapter 13
PLANNING YOUR CAREER   301

Appendix
ETHICAL STANDARDS OF HUMAN
SERVICE PROFESSIONALS   327

Index   333

Preface

To The Instructor

Welcome to the second edition of *The Human Services Internship: Getting the Most from Your Experience*. You will note a number of changes in this edition. Among these, two new chapters focusing on communicating with clients and career planning have been added. Attention has been directed toward websites where students may continue to explore various topics. A number of chapters have been strengthened through the addition of more detailed content and current information. Exercises in each chapter have been streamlined and categorized as to the nature of student thinking required in the task.

You will also note that much in the book has remained the same. As with the first edition, the book is designed for use by students and instructors who are engaged in the challenging experience of fieldwork in human service agencies. The text can be used successfully with students at any level in their educational program, by majors and non-majors alike. Although it is written for students taking course work in human service education programs, it may be used to great advantage with any student engaged in fieldwork in a human service agency, including those in psychology, sociology, and other social science disciplines.

Because the structure and organization of field experiences vary broadly from program to program, the text is written with this is mind. Students will benefit from using it in course-linked fieldwork projects, one-month "mini-term" experiences or longer half-semester, full-semester, or full-year practica and internships. The text is ideal for use in field seminar courses where various exercises and topics can be further discussed, but it may also be used fairly independently by students.

Purpose

Students in their human service fieldwork are expected to meet multiple and complex objectives. The purpose of this text is to provide the information, structure, guidance, and coaching that they need in order to approach their work with greater confidence. The goal of the text is to help students maximize their learning in every experience. The text is realistic, practical, and supportive in its approach.

During their fieldwork, students are supervised closely by agency supervisors but may have relatively little direct contact with faculty members and classmates as compared to their experiences in traditional classroom-based courses. Although fieldwork students may be placed in rich learning environments and work under good supervision, the demands of the workday often preclude the opportunity for them to discuss and reflect on their experiences on a daily basis. As a result, opportunities for learning all that they might from various experiences are sometimes missed. As an instructor of field courses, I have often wished that I could sit down with each of my students at the end of the day to discuss their experiences, call attention to important issues, raise questions, and help them to draw upon the knowledge base of the profession to make sense of their experiences. Obviously, this is not possible. This text engages students in a similar process of thinking and reflection.

Another common concern among faculty teaching field courses is how to help students make meaningful connections between their classroom learning and their field experiences. This text offers specific methods and exercises that students and instructors can use to accomplish this goal more effectively. An entire chapter is devoted to teaching students a six-step model for processing their experiences, a model that includes the application of theoretical and conceptual knowledge, as well as other types of reflection. Also, the exercises included in each chapter take the student beyond the level of learning content and into critical thinking and active application of content to their own experiences.

In this second edition, the exercises have been classified into one of three types: Personal Reflection: Observation of Self and Others; Synthesis: Linking Knowledge and Experience; or Analysis. This categorization is an effort to highlight for students and faculty the nature of the thought process required in the exercise. Each chapter includes several exercises, but there is no expectation that students will complete all of the exercises in each chapter. Faculty are encouraged to select and assign the exercises that will best help their students achieve the learning goals of their academic program. Faculty also might consider asking students to jot brief notes in response to some of the exercises while requiring more thorough completion of others.

Added to this second edition is an exercise at the end of each chapter labeled "For Your E-Portfolio." Electronic portfolios are explained in some detail in Chapter 1. This feature has been added in response to the increasing use of electronic portfolios in academic programs to assess student work and to encourage student reflection on their own work. The use of e-portfolios has become particularly prevalent in assessing student learning outcomes at the end of an academic program. Because the internship is the capstone experience in many human service programs, it is a likely point for such assessment. Students also value the building of portfolios to highlight their best work and sometimes use these portfolios in their job searches. The e-portfolio prompt at the end of each chapter is designed to facilitate the development of this work. Each prompt requires students to respond to the topic of the chapter and relate it to their own development in the internship. Faculty should, of course, feel free to adapt these prompts to the assessment processes and desired student learning outcomes of their own academic programs.

Philosophy and Approach to Experiential Learning

The text reflects certain assumptions about experiential education, about the role of writing in thinking, and about how students learn. The text assumes that while experiential education can be a powerful pedagogical approach, students often need to *learn how to learn* most effectively from experience. Although experience can be a good teacher, at times it may not teach very effectively in and of itself. Without careful thought, analysis, and reflection, students can easily draw erroneous conclusions and make incorrect inferences based on their experiences, just as they can misread or misinterpret a text. Experience is a powerful teacher that proves to be more effective when combined carefully with critical thinking, self-evaluation, and reflection. This text is designed to keep students engaged in this thinking process throughout their fieldwork.

Furthermore, the text requires students "to think on paper." This feature is based on the premise that writing helps the thinker to think more clearly and precisely. In each chapter, students express their thoughts in writing in response to specific exercises. As students' thoughts are expressed in writing and shared with the instructor, a fruitful dialogue can be initiated. Through reacting to the student's written work, the instructor has regular opportunities to provide supportive assistance, corrective feedback, prompts toward further reflection, or other responses that can enhance the student's learning.

Finally, the text is based on a philosophy of active, student-centered learning. Each chapter actively engages students by consistently bringing their experiences and thoughts into the discussion along with theoretical and academic content. Also, the inclusion of student work enlivens the text by offering concrete examples of student experiences and reflections.

Content

As in the first edition, much of the text's first three chapters deal with practical matters in beginning the field experience, focusing on issues such as myths about internships, stages in the development of internships (Chapter 1), development of a learning agreement (Chapter 2), and getting to know the field placement agency (Chapter 3).

Chapter 4 deals with helping the student understand more about the processes involved in learning from experience and introduces a six-step model that students can use in thinking through their experiences in the field. The model is my original work, building upon earlier models of experiential learning, and is a useful framework for helping students to extract maximum learning from their experiences. In summary, this model calls upon students to be careful observers of their experiences, to reflect upon personal reactions, to identify and apply relevant knowledge, to identify dissonance (i.e., points of discomfort or conflict that might include ethical dilemmas, conflicts between theoretical points of view, etc.), to articulate their learning from the experience, and to make plans for the next step in their work and in their learning. This process not only helps students learn more during their fieldwork but also teaches them a method for thinking through their experiences that can serve them well throughout their careers.

Chapters 5 through 10 deal with special skills of human services professionals. Chapter 6, Communicating with Clients, is new to this second edition, reflecting the importance of students' development of communication skills in working with individuals, families, groups, and communities during their internships. As in the previous edition, there are also chapters devoted to learning to use supervision (Chapter 5), dealing with diversity (Chapter 7), developing ethical competence (Chapter 8), preparing oral and written reports (Chapter 9), and dealing with emotions and stress in the workplace (Chapter 10).

Chapter 11, Trouble-Shooting, serves to raise students' awareness of some potential problems in human services internships and to help students avoid them. The chapter covers such topics as maintaining personal and professional boundaries, guarding against dual relationships, and maintaining personal safety. The final chapters of the text move toward providing closure for the student as the internship draws to a conclusion. Chapter 12 focuses on processes involved in ending the internship, including both evaluation and termination. Chapter 13, new to the second edition, assists students in making the transition into "life after the internship" through a focus on career planning.

In some cases, the chapters have been ordered in a way that reflects the stages of the internship's development. Chapters 1 through 3, for example, focus on getting a good start in the internship, whereas Chapters 12 and 13 focus on bringing it to a satisfactory close. Other chapters, such as Chapter 7, Dealing with Diversity, and Chapter 8, Developing Ethical Competence, are less clear in terms of their logical placement in this order. In making decisions about how to order the chapters for their maximum effectiveness, instructors should give thought to the objectives of the specific field experience in which their students are engaged and the emerging readiness of students to examine various issues. For example, Chapter 11, Trouble-Shooting, though placed relatively late in the text, might be assigned earlier in many situations due to its focus on preventing some common problems and pitfalls in fieldwork.

Throughout this book, content is included in the form of case studies and examples of various kinds. This material was inspired by nearly 25 years of experience in working with students and clients in various contexts. In no case, however, does this book include information drawn from any particular student or client. The people with whom I have worked over the years have provided a rich history upon which to draw, but their struggles and situations appear in this book in very disguised, composite forms. Names, identities, situations, and details have been routinely altered to such an extent that no actual person or persons are depicted. Any likeness to the names and circumstances of real people is strictly coincidental.

Acknowledgments

I wish to acknowledge gratefully the contributions of many people who contributed to the successful creation of this second edition, including the following: my husband and children (Rick, Alison, and Emily), who were unfailingly patient, encouraging, and supportive as I spent many long hours at the computer. I especially thank my daughter, Emily Kiser, for help as a research assistant and proofreader. Many thanks must also go to faculty colleagues and administrators at Elon University, who provided interest in and enthusiasm for my work; my students at Elon University who have taught me much of what I know about experiential education; my colleagues in the Southern Organization for Human Service Education and the National Organization for Human Service Education who encouraged my efforts; and Alma Dea Michelena, editor for this second edition, whose suggestions, guidance, and support did much to strengthen the book. Also I would like to acknowledge the work of Lisa Gebo, former Executive Editor at Wadsworth, whose support over the years I have valued highly. The staff of Matrix Productions also provided excellent assistance in strengthening the final manuscript and moving the book into production. I especially thank Aaron Downey and Ann Whetstone for their contributions to this second edition. Finally, I wish to acknowledge the time and contributions of the following reviewers who made astute and insightful suggestions, adding significantly to the quality of the finished work: Alisabeth Buck, Tacoma Community College; Mary Di Giovanni, Northern Essex Community College; John Hancock, Fitchburg State College; Jeffrey Haber, Metropolitan State College of Denver; Paul Hand, Anna Maria College; Ed Neukrug, Old Dominion University; Theresa A. Bowman Downing, Thomas Edison State College; Mary Kay Kreider, St. Louis Community College–Meramec; Lynn McKinney, University of Rhode Island, for the first edition. For the second: Kathleen Conway, Wayne State College; Anita Vaillancourt, University of Northern British Columbia; and Keith Willis, Wayne State College.

Introduction

One of the best ways to learn about human services and prepare yourself for a human service career is through fieldwork—that is, actually spending time in a human service agency engaged in its daily work. Most graduates of human service programs report that their field experiences were among the most important, valuable, and enjoyable parts of their professional preparation. As a student in a human service program, you have probably participated in other types of experiential education, as well because learning from experience has always been heavily emphasized in human service education. Since human service education began in the 1970s, students have participated in simulations and case study analysis, group discussions have revolved around student field experiences, and extensive fieldwork has been required for degree completion. *The National Standards for Human Service Worker Education and Training* (Council for Standards in Human Service Education and Training [CSHSE], 2005) defines field experience as "a process of experiential learning that integrates the knowledge, theory, skills, and professional behaviors that are concurrently being taught in the classroom" (CSHSE, 2005).

Human service programs vary in the structure of their field component. *The National Standards for Human Service Worker Education and Training* says, "While there is agreement that field experience is a critical component of any human service training program, there are variations in format, duration, and placement of the field experience" (CSHSE, 2005). There are indeed many formats for human service field programs, but you will probably find that your department's field program includes some combination of the following components:

1. Students engage in fieldwork early in their academic program while enrolled in one or more related academic courses. This type of experience is often referred to as a *field practicum* or perhaps as a *service-learning project*.
2. Students engage in fieldwork at approximately the midpoint in the program, after having studied human service content in traditional academic courses. Field experience at this point is especially useful as it enables students to apply their previous theoretical learning to their fieldwork and then return to coursework with greater understanding and insight based on direct experience in the field.
3. Students engage in fieldwork at or near the end of their academic program. This experience, often referred to as an *internship*, serves as a capstone experience, allowing students the opportunity to apply and test what they have learned in the classroom, as well as an opportunity to gain new knowledge and skills.

Despite the fact that human service students are nurtured and developed within such a tradition of active, experiential learning throughout their course of study, you may find that you are approaching your fieldwork with a sense of anxiety, a fear of not being sufficiently prepared. As one student said, "What if I go into my internship and find out I don't know anything? What if I fail?" Although this anxiety is an understandable and normal part of starting a new experience, in most cases there is little basis for this fear in reality. The foregoing discussion illustrates that students entering a field experience at any point during their human service program are not cast into the practice world without knowledge, information, or preparation for the task at hand.

Perhaps the greatest challenge for students in a human service field experience is not one of *possessing* knowledge but one of *making use of* that knowledge in practical ways. The field component should be central to your learning, serving as an integrative experience in which you retrieve your previous learning and apply it to practice situations. This objective can be quite daunting, especially in view of the broad curriculum that you have probably studied within your human services program. Content in human service education covers a wide range of topics and skills, including, for example, information about special populations and human problems; theories of human behavior and human development; skills in working with individuals, families, groups, and organizations; professional ethics; multicultural awareness and skills; and development of self-awareness. Despite its challenges, your task during your fieldwork is to make use of this material (and more), applying it accurately and skillfully to practice situations that you will encounter in the field. Through your field experience, your academic learning can come alive and take on new meaning as you see the connections between the knowledge and skills you have gained in the classroom and your "real-world" practice experiences while working in a human service agency. This dynamic connection between the academic and the practical makes fieldwork experiences in human services especially satisfying, interesting, and challenging.

With all of this in mind, the goals of this text are (1) to help you integrate theoretical and conceptual information with your experiences in the field, (2) to help you learn more from your experiences in the field by thinking extensively and carefully about those experiences, and (3) to provide the information, structure, and coaching necessary for you to explore the relatively unfamiliar territory of a fieldwork experience with confidence. Toward these ends, you will find that the text includes useful information about every stage of the process from preparing for the experience (Chapter 1, Getting Ready) to ending the experience and moving on (Chapter 12, Ending Your Internship and Chapter 13, Planning Your Career). Each chapter calls upon you to be an active learner, reacting to, applying, and reflecting upon the many ideas discussed. As you work through the material and exercises in the text, they will guide and support you through the various stages of your field experience, helping you to seize its opportunities, anticipate and avoid its potential pitfalls, and extract maximum learning from your experiences.

The human service literature uses a variety of labels to refer to field experiences in human services. The terms *practicum*, *fieldwork*, and *internship* are frequently used to denote various types of field experiences. Nomenclature varies from program to program and has been the subject of much discussion in human services programs in recent years (Simon, 1999). For the sake of clarity, *fieldwork, field experience,* and *internship* are the terms generally used in this book, and they are used interchangeably.

Now as we embark on your fieldwork experience together, best wishes to you! Keep in mind what a privilege it is to be allowed the status of an "insider" within a professional organization. Be determined to use every opportunity to advance your learning and development as well as to serve others. These attitudes will set the stage for an unforgettable and invaluable learning experience.

References

Council on Standards for Human Service Education and Training. (2005, May). *National standards for human service education.* Retrieved Dec. 5, 2005, from http://www.cshse.org/standards.html.

Simon, E. (1999). Field practicum: Standards, criteria, supervision, and evaluation. In H. Harris & D. Maloney (Eds.), *Human services: Contemporary issues and trends* (2nd ed., pp. 79–96). Boston: Allyn & Bacon.

Chapter 1

Getting Ready

If this is your first fieldwork experience, you probably have many questions about it. You might feel uncertain about what to expect as well as unclear about what is expected of you. As a result of this ambiguity, you might find that you have some mixed emotions about doing an internship, ranging from anxiety and dread to excitement and eagerness. This chapter will help you to feel more comfortable as you approach your internship by giving you a clearer idea of what an internship is (and isn't) and the types of learning goals it is meant to accomplish. In order to bring the internship into clearer focus, the chapter considers some common myths about internships as well as faculty and student perspectives on why internships are valuable.

A Student's Reflections on Preparing for Internship

I can't wait to do my internship! I have been looking forward to this for two years. I must admit that I do have some worries and lots of questions about the internship though. For example, how do I go about getting an internship placement? Once I have a placement, how am I supposed to fit an internship into my life? I don't have time for my family and friends even now! What will be expected of me? Will I be like an employee? A volunteer? How much will I work? All day? Every day? Will I work the same hours each day? It seems that I have more questions than answers at

continued

A Student's Reflections on Preparing for Internship *continued*

this point, but I am looking forward to it anyway. One thing I do know is that I will get some real-life experience in the field that I hope to work in after college. I appreciate the opportunity to do that.

So What Is an Internship?

It can be difficult to form a clear and cohesive picture of what an internship is like, especially if you have never participated in such an experience before. One of the factors that contributes to this difficulty is the fact that internships can take many forms. For example, in some programs, the time that students spend in the internship is like a full-time job. Students work full days every day in their placements, and their curriculum is arranged so that no other coursework interrupts the day. Such internships are generally referred to as "block placements." Other programs use "concurrent placements," meaning that students perform their internships while also being enrolled in traditional classroom courses. In such arrangements, students attend their placements on alternating days or establish other types of part-time schedule arrangements.

What makes an internship worthy of the name is not how the time is arranged or whether it is part-time or full-time, but the total amount of time spent in the placement, how that time is used, and the degree of reflection, application of previous learning, and other forms of critical thinking that are included in the experience. A number of definitions of the term *internship* offered in the literature touch upon such issues. The Council for Standards in Human Service Education uses the terms *fieldwork* and *internship* synonymously and defines these terms as:

> the advanced or culminating agency-based experience which occurs toward the end of a student college/university experience. This usually requires close supervision from agency personnel and the college/university faculty including regularly scheduled seminars with a faculty member. The experience of fieldwork provides a bridge between the academic experience and later professional employment. (1995, p. 21)

Cantor echoes the idea that the internship serves to link the academic experience with the professional world when he says, "Internships are experiential learning activities that bridge academia and the profession for students" (1995, p. 71). The same theme is touched upon in the following definition, which also reflects something of the multifaceted nature of internships as well: "Part audition, part test, part question, part answer. Lasting for a summer or during the school year, for a semester, interns are usually college students trying out their career by working in a professional (or near-professional) capacity" (Fry, 1988, p. 7).

The following description, though lengthy, is clear enough to be helpful and broad enough to be inclusive of most field experiences in human services: The human service internship is an experience in which a student, sponsored by an educational institution, engages in education and training while working in a human service organization or role for a substantive period of time. The student engages in an ongoing process of observation, practice, and reflection in order to learn from experience. From so doing, the student gains new knowledge and skills, applies academic knowledge to practical situations, refines previously acquired skills, and gains a greater understanding of self and the human services profession.

To gather more information specifically about human services internships, you might visit the website for the Council for Standards in Human Service Education. There you will find suggested parameters and guidelines for internship experiences that ensure the quality of student learning.

A Student's Reflections on the Internship

At my internship I am treated as a regular staff member and not as an intern most of the time. In fact, I sometimes forget that I am still a college student. It's great to be treated as an equal and to know that my opinions are valued by others. I have the same responsibilities as the staff when it comes to dealing with the residents. As an intern though, I have been given a broader experience so that I can learn more. I not only work with the residents but with their families as well, so part of my day is spent with the family workers. The staff is letting me see the whole operation and not just one aspect. Best of all, I am gaining confidence every day, and I'm learning things about myself that I never knew before.

Myths About Internships

Even with clearer definitions of what an internship is, students can still enter the experience with some unrealistic expectations. Under such circumstances, it is inevitable that students will encounter problems, disappointments, and misunderstandings. Therefore, an essential step in getting ready for your internship is clearing up the myths and misconceptions that many students hold about internships. As with most situations in life, you will experience far more satisfaction and less frustration if you understand beforehand what is realistic and unrealistic to expect in the situation.

Myth #1: My Fieldwork Will Probably Lead to a Job Within That Agency

Although some internships do extend directly into employment for the student in the host organization, this outcome is more the exception than the rule. Nevertheless, it is important to recognize that your experience there can, and probably will, make you far more employable. Not only will your skills be enhanced by your experience, but your resume, your references, and the quality of your job interview responses will improve as a result of your field experience. Also, internships provide numerous networking opportunities in which you can meet professionals from other agencies within the human service system. These contacts can be a great asset when you conduct your search for a position. It is fair to expect that your field experience will help you secure a job, but in most programs relatively few, very fortunate students are actually employed by their agency immediately following the field experience.

Myth #2: While Doing My Fieldwork, I Will Learn by Doing and Will Not Have Assignments and Homework as in Traditional Classes

Although you will no doubt learn by doing, this is not the only manner in which learning takes place during your fieldwork. Assignments vary from program to program and from professor to professor, but typically students are required to write and read extensively during their field placements. Students are generally required to write reflection papers and/or daily journal entries about their experiences, research papers about the problems and populations with which they are working, and/or book and article reviews based on a reading list developed for the internship. Additionally, students attend class regularly (often weekly), meeting with other students in a seminar-style setting to discuss their experiences. Final grades are usually based on the quality of students' work on academic assignments as well as the quality of their work in the agency.

Myth #3: An Internship Is Like an Apprenticeship. I Will Work Under a Qualified Professional and Learn to Do as He or She Does

An apprentice is generally thought of as an individual who is learning technical skills from a more experienced and skilled person. Although it is true that you will be working under an experienced supervisor and will have the opportunity to observe that supervisor and others as potential role models, the learning involved is more complex than the above statement implies. Human service work is sometimes described as both an art and a science (Gladding, 1992; Young, 2005). This refers to the fact that although some tasks may follow a clear protocol of how to proceed, many tasks in human services rely to some extent upon personal style and judgment as to how and when they are done. Your field experience will require that you observe, analyze, critique, and evaluate yourself and others. Out of this effort your own personal style of helping people and working with others will emerge. Although you will probably emulate various role models as you develop your own professional style, the critical thinking and independence required of a developing human service professional go well beyond that of an apprentice.

Myth #4: As an Intern I Will Have the Responsibilities and Autonomy of a Professional Staff Person

After years of sitting in classrooms, you are probably eager to be active and highly involved in your agency. It is sometimes frustrating to find that as the experience begins you are asked to observe others, read related materials, and observe meetings in which you may be able to participate very little. This period is a predictable stage in the development of the internship. You need time to become acclimated to the agency and learn basic procedures and policies. This period also allows your supervisor and other staff members to become familiar with you and to become comfortable with your capabilities.

As the placement progresses over time, students are typically allowed more autonomy. The degree of responsibility and autonomy that a given student is allowed varies considerably depending upon the nature of the agency, liability and legal constraints within the agency, the student's level of personal and professional skills, and the field instructor's supervisory style. Students should not expect to assume the role of a professional automatically but should recognize that, depending upon the particular situation, this status may be earned as the field experience progresses.

Myth #5: Doing an Internship Is Like Being a Volunteer in the Organization

Because most students have been volunteers before but have not been interns, they sometimes enter their fieldwork agencies assuming that there are similar expectations. Like volunteers, most students are not paid for their work during the field experience, but the similarity between volunteer work and fieldwork ends here. Volunteers are generally asked to do whatever the agency needs to have done, whether this is direct work with clients, clerical tasks, or even building and grounds maintenance. In contrast, the central focus of a fieldwork experience is the student's learning rather than the agency's needs. The activities of a student should be identified and planned based primarily on their educational value. In most field placements, as in most jobs, the student will perform a variety of tasks, some with greater educational value than others. A quality field experience, however, will engage the majority of the student's time in activities that develop and refine professional knowledge and skills. If you should find that the majority of your time is being spent on tasks with low educational value, you should discuss this with your field supervisor and with your faculty liaison.

A further distinction between a volunteer and an intern is that a greater level of responsibility is generally expected of an intern. Because interns are preparing for professional careers in human services, they are expected to adhere to professional standards in their work. Professional dress, reliability, promptness, and other characteristics of professionals are clearly expected of interns. Volunteers, on the other hand, may be granted more latitude in such matters. As distinct from volunteers, interns generally enjoy certain privileges as well. Most interns, for example, have access to client records and participate in such professional-level activities as staff meetings, case conferences, and in-service training sessions.

Myth #6: Doing an Internship Is Like Doing a Service-Learning Project

In recent years, service-learning has become a growing trend on college campuses. In service-learning experiences, students participate in community work that links with the learning objectives of a particular course. Because all academic disciplines engage in service-learning, students might participate in service-learning projects in classes such as history, literature, and biology, as well as in human services classes. Human service internships are similar to service-learning projects in that there is a focus on students learning from work in the community.

There are also, however, some important differences between these two types of learning experiences that are worth exploring. Service-learning projects place a strong emphasis on meeting a community need. While student learning is also a goal, this goal is balanced with the need to serve the community. In service-learning, there is also a strong focus on the principle of reciprocity, that is, the community must benefit as well as the student. In contrast, human service internships are primarily focused on the student and student learning. While some internships might also provide an important service to the community, this is not a necessary condition for internships. Some supervisors and/or agencies, for example, do look to interns to provide a specific service or to produce a particular "product" during the internship. Therefore, some internships may be thought of as service-learning internships while others may not.

Both types are equally valuable as internships, and students in human service programs typically have many experiences in which the focus is on service. Learning to serve is obviously an important goal of all human service programs. Nevertheless, service is not the primary goal of the internship. In order to prepare students to perform as professionals, the primary goal should clearly be on student learning.

Myth #7: If I Do Not Enjoy My Internship, I Must Have Chosen the Wrong Field Site or the Wrong Career

Just as with classroom courses, some of the most valuable field experiences are those that push you outside your comfort zone. This means that it is necessary to be uncomfortable at times to experience the greatest personal and professional growth. Enjoyment should not be the measure of your satisfaction with your placement. Rather, it is more useful to examine how much you are being challenged, both personally and professionally, by the environment and experiences that it has to offer.

It is also fair to say that a single field experience in one organization cannot fully reflect what working in that field is like. Many factors play into the quality of your experience as an intern, including the professionalism of the staff, the quality of the agency's programs, the organization's culture, the nature of the organization's relationships with clients and the larger community, and the quality of the leadership and management of the organization. It is unfortunate when the dynamics of a specific organization lead a student to conclude that an entire field of service or client population is not for them. If you do not have a positive experience in your internship, try to sort out for yourself the specific factors that are problematic and try not to generalize to an entire population of clients or type of organization.

Myth #8: For My Fieldwork to Be Most Useful, It Should Be in the Same Type of Job I Would Like to Find After Graduation

Although this statement seems to be common sense, it is not always true. Although it is important to keep your career goal in view, it is often not possible or even advisable to enter a field placement in your desired position. There are usually a number of developmental steps to be taken to achieve professional competence in a given role. Your fieldwork can often best be used to help you achieve one or even several of these developmental steps.

For example, a student who was interested in becoming a family counselor was disappointed to find that she could not secure a placement in this field due to concerns of privacy and confidentiality, as well as her level of training. In her human service program, three field experiences were required. For these three opportunities, she was placed in a day-care center, a battered women's shelter, and a group home for emotionally and behaviorally disturbed children. Although none of these placements offered her the exact role to which she aspired, each of them contributed knowledge and skills that were important to her overall preparation for the role of family counselor.

Myth #9: Everyone Gets an "A" in the Internship

Grades in fieldwork courses often follow a distribution more similar to classroom courses than many students expect. As discussed under Myth #2, a number of different types of assignments are factored into the internship grade. There is often wide variance in student performance levels in the field sites, ranging from students who perform exceptionally well to those who perform adequately to those who perform below average or even unsatisfactorily.

Before arriving for the first day of your internship, you would be well advised to read the course syllabus carefully in order to understand the various assignments and requirements of your fieldwork experience and to understand how the different components will be weighted in assigning your final grade. Also, you should pay special attention to the items on the evaluation instrument that your field supervisor will use to evaluate your performance. Becoming familiar with these documents will clarify the exact expectations and standards that you need to meet in order to earn a grade that you can be proud of.

Why Do an Internship?

The discussion in this chapter has set the stage for your thinking more clearly about your own internship and the expectations you have for it. Now that you understand the general parameters of the field experience, it is time to bring your own internship into clearer focus. The following exercise asks you to think about and describe your own hopes and expectations for this experience.

◆ EXERCISE 1.1 PERSONAL REFLECTION: OBSERVATION OF SELF AND OTHERS

Before embarking on any journey, it is wise to clarify exactly why you want to make the trip. Why do you want to do an internship? What do you hope to gain from this experience? Take a moment to think about this and record your thoughts.

Each person has a different set of experiences, circumstances, and expectations influencing why she or he wants to do an internship. As a result, responses to the question "Why do fieldwork?" will vary broadly. Because academic programs in human services invariably require field experiences, faculty members must also answer this question. What rationale do faculty members put forth to support the fieldwork requirement? Some common rationales are summarized next. Examine these and compare them to your own earlier responses.

- ◆ Fieldwork helps to identify and modify student attitudes that may interfere with effective practice (Dore, Epstein, & Herrerias, 1992).
- ◆ Fieldwork helps to develop certain personality traits in the student that are conducive to effective practice (Dore et al., 1992).
- ◆ Classroom learning, even with extensive use of simulations and case studies, is not sufficient to produce effective human services professionals. Students need an opportunity to practice their skills in the real world.
- ◆ Fieldwork allows students an opportunity to apply and test classroom learning in practical, concrete situations. Students can learn to develop and test hypotheses based on theories, concepts, and knowledge that they have gained in the classroom.
- ◆ Professionalism can best be taught in the field as students have an opportunity to observe professional models over a significant period of time and think critically about their work. Students also learn to evaluate their own work and set goals for further development as professionals.
- ◆ Students can learn more about a particular subfield within human services that they might wish to enter. This direct experience can help students dispel false impressions they might hold about the field as well as gain additional skills or knowledge needed to work effectively in that field.
- ◆ Fieldwork allows students to integrate and synthesize learning that has been fragmented in the academic curriculum. Knowledge, skills, attitudes, and values must be unified and used in conjunction with one another rather than treated as distinct entities.
- ◆ Students need an opportunity to make a transition from the academic world to the working world. The habits formed within the academic environment are sometimes not those conducive to success in the working world.

◆ Fieldwork encourages personal growth in ways that differ from the classroom. The active learning, risk-taking, decision making, and personal responsibility that are inherent in agency-based practice are conducive to growth and developing professionalism.

Understanding Your Internship as Developing Over Time

Although you might think about your internship as one unified experience, you will more likely find that it develops and changes over time, somewhat like a living organism. Nothing is born fully developed, including your internship, so it is helpful and realistic to expect that your fieldwork experience will change as time goes by.

Many models have been suggested for understanding and describing the development of internships, with little evidence of any one model's accuracy as compared to that of others (Diambra, Cole-Zakrzewski, & Booher, 2004). For example, research on music therapy internships (Grant & McCarty, 1990) and psychology internships (Lamb, Barker, Jennings, & Yarris, 1982) reveal evidence for a five-stage model, consisting of pre-entry preparation, finding a place in the organization, gaining identity and role differentiation, emergence of professional competence and independence, and separating. Inkster and Ross (1998) describe a similar six-stage model of internships: arranging the internship, orientation, reconciling expectations with reality, productivity and independence, closure, and reentry. Sweitzer and King (2004) present a five-stage developmental model for the internship that includes anticipation, disillusionment, confrontation, competence, and culmination.

Some writers have suggested that human services internships progress through stages similar to those of the helping process itself, including such stages as exploration, relationship building, and goal setting (Grossman, Levine-Jordano, & Shearer, 1991; Siporin, 1982), while others have suggested stages of internship development that parallel the eight stages of human development described by Erik Erikson, including stages such as developing trust, establishing autonomy, and developing initiative (Kerson, 1994). A number of writers have described the development of internships in a far more general way, as a series of tasks and issues that shift in focus as the internship moves from beginning, to middle, to end (Chiaferi & Griffin, 1997; Michelsen, 1994; Pettes, 1979).

Clearly, there is much agreement that internships do progress through certain developmental stages but little agreement as to the exact nature of these stages. The four stages of internship development discussed in the subsequent section of this chapter reflect a synthesis of ideas from a number of different models in conjunction with my own observations from working with numerous interns. Understanding some of the predictable ways in which your internship is likely to develop and change over time will better prepare you to master the challenges of each stage.

Stage 1: Preplacement Stage

The first stage of the internship's development occurs before you arrive for the first day of work. This first stage is critical to the success of the placement because important foundations for your work are laid during this time. Usually the student and the faculty members of the human service program work together to decide upon an appropriate field placement. Efforts are made by you and/or faculty members to secure a placement compatible with your learning needs. Once a positive contact has been made with an organization, you will make an initial visit for an interview with your field supervisor. Both parties, you and your potential supervisor, form initial impressions of one another during this time. Your supervisor will be particularly interested in forming an initial impression of your interests and learning needs.

If you both see the placement as appropriate and workable, the agency begins to make preparations for your fieldwork. Your field supervisor will plan for your arrival by giving thought to such issues as the types of assignments to give you and the location of a workspace for you. Colleagues, and in some cases even clients, will be informed and prepared for your impending arrival. As you approach the internship, you too will be making necessary preparations. Clearing your schedule of any responsibilities that could interfere with your internship, engaging in any reading that might prepare you for your responsibilities, as well as more mundane activities, such as pulling together an appropriate wardrobe, are all essential parts of your personal preparation. (The discussion and exercises in Chapter 2, Getting Started, will help you to work through the steps involved in this stage of the internship successfully.)

Stage 2: Initiation Stage

The first days and weeks of an internship might best be described as an orientation and a series of multiple beginnings. As you meet new people and become acquainted with the agency, you will probably experience a range of emotions, including excitement, anxiety, and anticipation. With this emotional intensity can come a burst of energy and eagerness to dig in and get started on the work of learning by doing. In contrast, as discussed earlier, supervisors and agency administrators may take a more measured approach to starting the internship. There is often reading and observation that they wish you to do before taking on any direct responsibilities.

In the initiation stage, your supervisor will also try to assess more fully your strengths and limitations in order to make appropriate assignments for you. Although students vary in their readiness for the internship and working independently, many express a need for structure and direction during the initiation stage of the internship. An orientation to the agency, a structured schedule of meetings that you might attend, and a set of tasks that you are to accomplish are particularly helpful during this time. If you find that your needs for structure and focus are not being met, tactfully approach your supervisor to discuss this.

Just as your internship will develop and change over time, your relationship with your supervisor will do so as well. During the early days and weeks of your relationship, a critical task is developing mutual trust. Direct, open communication is essential to building this trust. The process of establishing your learning agreement (discussed in detail in Chapter 2) is one of the most important tasks of the initiation stage and is an excellent vehicle for establishing open communication with your supervisor about your learning interests and needs in the internship. Also essential in building trust is proving your own trustworthiness by being reliable, prompt, and eager to assume responsibility.

Although each stage of the internship offers unique opportunities for you to grow and develop your skills, each stage can offer potential pitfalls as well. A brief discussion here of these pitfalls will better equip you to avoid them. One common pitfall of the initiation stage is a tendency to "rush to judgment" about the quality of a placement. This judgment may be positive, launching you into a honeymoon phase in which you believe you have the most perfect internship and supervisor that any student could ever have. This point of view may be wonderful while it lasts, but all too often ends painfully later when this idealized view of the situation is marred by the realities of everyday demands in the organization. The student who is overly optimistic about the internship might not be sufficiently careful and assertive in establishing a clear learning agreement, feeling unduly confident that the experience is going to be ideal. As a result, the learning contract might not be clear enough or provide sufficient challenge to ensure that the student has a quality experience throughout the internship.

Of course, a negative rush to judgment is equally problematic, if not more so. Negative judgments can understandably result when, for example, a student enters the organization on a particularly chaotic day, is treated rudely by staff, or is ignored. At times, student expectations of an organization and that organization's reality can be so far apart that major adjustments are required of the student. If you should find this to be the case in your internship, you will no doubt feel both frustrated and disappointed. In such a situation, try to keep your attention focused on two central questions, "What can I learn here?" and "How can I learn it?" All organizations have their strengths and weaknesses. Most students have much that they can learn, even in less-than-ideal settings. If your field site has some problems, changing this situation may be largely outside of your control. What is within your control is your own level of motivation and determination to learn. In the end, these characteristics will be at least as important to your learning as are the qualities and characteristics of the setting in which you are placed.

In a similar vein, Chiaferi and Griffin point out that "all too often, students focus on what they deem to be the shortcomings or personality defects of their supervisors while failing to recognize their own contribution to difficult interactions" (1997, p. 26). This observation suggests that students would do well to work on their own self-awareness during this initiation stage. Although it is inevitable and appropriate that you evaluate the quality of your agency and your supervisor, it is even more important to evaluate your own behavior and attitudes continuously, reflecting upon how they are affecting your ability to establish a sound foundation for your internship.

Stage 3: Working Stage

As relationships are developed, the learning agreement completed, and expectations voiced and negotiated, the internship begins to move into a third stage, which might be referred to as the working stage. During this stage, less energy is invested in establishing relationships and negotiating mutual expectations and more energy is invested in achieving your learning goals and in accomplishing the agency's work. The trust that has been developed between you and your supervisor by this stage generally results in your supervisor allowing you more autonomy and allocating more responsibility to you. Also, by this time in the internship, both you and your supervisor have achieved a better understanding of your strengths and your learning needs. In fact, an early signal that your internship is entering this stage may be that you find yourself gaining comfort in asking questions and admitting gaps in your knowledge. This important shift enables deeper, more meaningful learning to occur.

Another characteristic of the working stage of the internship is that daily activities settle into more of a routine. You have acclimated to the agency culture and learned agency procedures. As a result, earlier anxieties are reduced and you may generally feel more relaxed and confident. This increased comfort and confidence, in conjunction with a greater understanding of the learning opportunities available within the organization and your supervisor's increased knowledge of your learning needs, might result in some renegotiation of your learning agreement at about midterm in your internship (Pettes, 1979). At this stage in the experience, both you and your field supervisor are in a better position to plan and implement learning experiences specific to your particular needs.

In the working stage of the internship, the student engages deeply in the process of learning to be a professional. Student learning during this stage focuses on such issues as developing practice skills, learning about agency administration, recording and other forms of written communication, developing professional responsibility, learning to use supervision, and learning to juggle multiple tasks effectively. As a result, the working stage can be an exciting, challenging, and deeply satisfying time in the internship.

A possible pitfall of the working stage is that students can lapse into carrying out routine duties, enjoying their greater comfort and confidence, without exploiting the learning opportunities that they might pursue. If the learning agreement has been too modest, you might even find that all or most of your learning goals have been accomplished at this early juncture. If this situation develops, you and your supervisor should revise the learning agreement to ensure that your time and activities are wisely invested.

A second potential pitfall of the working stage may be the tendency for students to begin working too independently. As an intern becomes more capable, the supervisor might tend to spend less time with the intern, resulting in less supervision, less teaching and, as a result, less learning. An intern in such a situation might be quite flattered, even relieved, to be more frequently away from the supervisor's watchful eye. Nevertheless, for the sake of your own learning, if this situation should develop, you should politely express your concern about the decreased supervision time to your supervisor.

> ### A Student's Reflections on the Internship's Development
>
> After four weeks on the job the new has worn off. Let's face it—it is no longer fun to get up at 6:00 A.M. every day. There's nothing wrong with my internship, but it has just become more routine now. My roommates are still on college student schedules, staying up till all hours. I'm falling into bed exhausted by 10:00 P.M. I must admit, though, that there are some rewards for hanging in there. I'm much more comfortable at my placement than I was at first, and I feel good that they trust me to do more on my own. My supervisor also told me that a lady who called and asked me some questions over the phone had commented to her about how polite and professional I was. It's nice to know that some of my strengths are beginning to show.

Stage 4: Termination Stage

The internship shifts into its final stage when the student and supervisor begin to anticipate and plan for the end of the internship. New assignments are not as freely given due to time constraints. The work that the intern is currently doing must either be completed, or plans must be made to transfer responsibility to other workers. As you bring the work of your internship to a close, you must say good-bye to supervisors, colleagues, and clients. Depending upon the nature of these relationships, this can be a time of great emotional intensity. At the same time, however, the termination stage can also be a period of equally great satisfaction and even pride.

The final stage of the internship is also a time for reflection. You might reflect upon such questions as: How well did I perform in this internship? What have I really accomplished here? What have I learned after all? What do I have regrets about? What do I value and appreciate about this internship? Where will I go from here in my professional development? Such reflection is a positive and valuable aspect of your experience and will help you to engage more meaningfully in the formal evaluation process with your field supervisor that occurs toward the end of this stage.

The quality of your experience during the termination stage will depend in part upon your own patterns and skills in dealing with endings. The termination stage will give you the ideal opportunity, of course, to examine and work on these patterns and skills. Therefore, many students find it useful for supervision during this stage to focus specifically on the skills involved in productive endings. Even though you might have become very competent and independent in handling various responsibilities during your internship, you might find that you need closer guidance and teaching in order to deal effectively with the new challenges of ending relationships.

As implied above, pitfalls during this stage tend to relate to the student's patterns in handling endings. Therefore, it is wise for students to evaluate these patterns prior to entering the termination stage. Some students have great difficulty letting go. These students might delay transferring tasks or clients to other workers, hanging on to these responsibilities as long as possible. Other students might tend to minimize the significance of the ending, repeatedly telling everyone, "I'll be back to visit." Denying that the end is coming can block the student from saying good-bye in a real and meaningful way and can even give clients the false reassurance that the student will still be available. Before making such casual promises, students must examine closely the appropriateness of such visits, their motives for wanting to stay in touch, the likelihood of following through with such promises, and the agency's policies about continued relationships with clients.

A third pattern that sometimes emerges as students approach the end of the internship might be described as "premature disengagement." As the student begins to anticipate leaving the organization, attention to the tasks at hand may decrease. Within this pattern, it is as though the student has emotionally and mentally left the internship before completing it. A student who has earlier been hardworking, motivated, and attentive to detail can lapse into daydreaming and distractibility as the end of the internship approaches. Obviously, this is a behavior pattern that can only harm the quality of a student's performance and the supervisor's evaluation of the student. Although it is challenging to work on letting go and staying engaged at the same time, this is exactly what the termination stage of the internship calls for. Throughout the termination stage, you should stay meaningfully engaged and dedicated to the people and tasks that are yours to respond to each day.

Positive termination is an important and challenging task that will need careful consideration and planning. Chapter 12, Ending Your Internship, will help you meet this challenge more effectively by dealing with termination issues in greater detail.

A Student's Reflections on Termination

Today was my last day. It was a good day but also depressing. This internship has been a lot to take in, but I learned so much. The people I worked with always took time to talk with me after group and other events to see what I thought and how I felt about what had happened. I appreciate the time they took with me and that they asked for my ideas instead of just telling me theirs. The internship also made me think a lot about my own issues and my family, and, even though I never discussed it with anyone, I feel that I was able to put into perspective some personal issues. This was an unexpected but wonderful benefit. Now for the depressing part. I have learned and grown so much that I hate to leave. I will really miss the people and feel just plain sad about the fact that it's over. I am thankful that I will have opportunities to do more fieldwork in this major in the future. That makes it easier to move on.

 ◆ **EXERCISE 1.2 SYNTHESIS: LINKING KNOWLEDGE AND EXPERIENCE**

Based on your previous fieldwork experiences, what stages of development have you observed in these experiences? Discuss your internship in terms of its development thus far. What particular developmental tasks have you already dealt with in your internship? Which ones are you currently working on? How do you see your experience thus far as fitting into the developmental stages of the internship described above? What particular pitfalls discussed in this section are you able to identify with as relevant to your experience thus far?

Developing an Electronic Portfolio

As you begin your internship, you might consider developing an electronic portfolio (e-portfolio) to document your experiences, your learning, and your reflections. In fact, some human services programs require students to develop such portfolios for the purpose of assessing student learning and growth.

Research on e-portfolios indicates that faculty members tend to see student portfolios as primarily an educational tool; that is, the portfolio is seen as a tool for fostering, documenting, evaluating, and measuring student growth and learning. In contrast, students tend to see portfolios as a marketing tool, that is, an archive that documents their skills and proficiencies to a given audience, especially employers, for the purposes of career development (Breault, 2000). Because this book focuses on the internship as an educational experience, the discussion of e-portfolios here assumes that e-portfolios are being developed as an educational tool. Nevertheless, it should be pointed out that good educational e-portfolios and career development portfolios are not mutually exclusive.

Educational E-Portfolios

Educational e-portfolios might be organized in various ways. In some cases, the human service program might have specific learning outcomes for students in the program and require that students archive documents to demonstrate their progress and increased mastery toward each outcome. In other cases, students might organize their portfolios around courses they have taken, submitting course syllabi, papers, exams, evaluations, and other materials. In some situations, students may have more latitude to develop their portfolios around broader themes such as knowledge, skills, values, philosophy of helping, and goals for professional development. Even if your human service program does not require you to construct a portfolio, you might consider doing so on your own, organizing it in a manner that would be most useful to you.

In some programs, students are required simply to compile their culminating work in the program in an e-portfolio. Such programs might require students to submit their responses to the "For Your E-Portfolio" exercises at the end of each chapter in this book. In other human service programs, students may be required to document their learning and growth from the beginning of the internship to its end. If this is the case, students will need to include in their e-portfolios some very recent samples of their work to document their starting point. In such cases, the program might require students to submit work from the end of the previous semester or from their last field experience prior to the internship. Alternatively, students might be required to begin the semester by writing a paper in which they assess their own mastery vis-à-vis each learning goal of their human service program. Later writings in the internship are then compared against this beginning document.

Still another model for e-portfolios is provided by those programs that require students to submit work throughout their program, including work from the earliest, beginning courses as well as from the advanced and capstone courses. Toward the end of the academic program, students are asked to review their work and comment on their own learning and growth (Barrett, 2000). This type of archiving and reflection can be powerful experiences for students as they recognize their own accomplishments. If your program has such a requirement, you probably have already started to build your e-portfolio and will add whatever documents are appropriate during your internship. You will then review all of these documents and reflect on your own learning and growth process toward the end of the internship experience.

This exercise is valuable because one of the primary goals of human services education is to produce reflective practitioners who are able not only to perform competently in the field but also to observe themselves accurately and see themselves as growing within their profession. E-portfolios are particularly effective in documenting this reflective process. The electronic nature of the portfolio offers unique options for students commenting on their own work through hyperlinks, electronic "sticky-notes," or other formatting and presentation options that are specific to digital media (Barrett, 2000).

E-Portfolio Benefits

Whatever the exact nature of your e-portfolio, creating one has many potential benefits in both its process and its product. As a process, it provides students a structure by which to examine and reflect on their own work and growth. As a product, it not only gives faculty the opportunity to measure student learning and growth but also to measure the student's capacity for self-observation, reflection, and analysis. Additionally, constructing an e-portfolio ensures that students develop and demonstrate computer skills that will enhance their effectiveness throughout their careers.

Having completed the portfolio in your internship, you will be well-equipped to develop a marketing portfolio to enhance your job search as you complete your academic program. The educational e-portfolio can serve as an excellent foundation or springboard for a career development portfolio because graduates clearly want to provide potential employers with evidence of many of the same competencies that their human service programs tried to cultivate. Baird (1999) suggests that developing a portfolio of your work as a student presents a real employment advantage. He asserts, "Unlike other students who can only report a grade point average and courses taken, you will have real, tangible evidence of your work product" (p. 8).

E-Portfolio Formats

Portfolios may be as simple or as elaborate as the creator desires. They may be as simple as papers three-hole punched and placed in a binder with appropriate subject dividers or they may be elaborate electronic files that include not only documents but multimedia products as well. Videos, PowerPoint® presentations, and even musical and artistic works can be submitted as appropriate, depending upon the software you use to store your information. Excellent portfolios with hypertext links can be produced using nothing more than up-to-date word processing software. Portfolios that include multiple media require software such as Microsoft FrontPage® or PowerPoint. It is important, however, to remember the portfolio's central purpose as a learning, reflection, and documentation tool. It is not wise to attempt such complicated technological tasks that you are overwhelmed by the technology or a reader is confused by its complexity.

Remember that the technology is simply a tool to help you document, reflect upon, and synthesize your learning, laying the foundation for your future growth. More information about e-portfolios is available from numerous websites such as The Reflect Initiative and The Electronic Portfolio Consortium. Each chapter of this book provides a focus on a specific topic that is central to the development of competent human service professionals. Therefore, the structure of the book can provide a structure for the construction of your e-portfolio. To facilitate your use of the book in this way, each chapter ends with a broad writing prompt to facilitate your reflection on that chapter's major topic as it relates to your development. Even if your program uses a different structure for student e-portfolios, the end-of-chapter e-portfolio reflections might be useful in a more general way in developing your thoughts about a given topic or skill area.

Conclusion

Clarifying expectations is an important step in getting your internship off to a good start. You, your field supervisor, and your faculty liaison have certain expectations of the internship. Knowing that your internship will develop through some predictable stages as well as being familiar with the issues and tasks inherent in each stage will better prepare you to meet the demands of the internship over time. Consider developing an e-portfolio over the course of your internship as a way of documenting your learning and experiences. Your supervising faculty member will be your best resource as to what to include in your e-portfolio as your internship begins.

As you prepare yourself for your internship, visit the website for the Human Services Career Network to get more information about how to get the most out of your internship.

◆ FOR YOUR E-PORTFOLIO

A portfolio for your human service program: Find out whether your human service program requires you to develop an e-portfolio. If so, learn all that you can about what you need to submit to the portfolio and how the portfolio is used within your program. Is it graded by faculty to determine how well you have achieved the learning outcomes of the program? Is it used to assess the human services program in evaluating how effectively it is meeting its learning objectives with students? How much flexibility do you have about the format and what you must include?

A portfolio for your own use: Think about what you would like to include in an e-portfolio. What materials would you like to archive for future reference? What examples of your own work from your human service program would you like to compare? For example, are there papers from your Introduction to Human Services class that you would like to compare to your writing and reflections during your internship? What knowledge and skills would you like to document and highlight for potential employers? How can you use your portfolio as a tool for lifelong learning? Are there examples of a professional's feedback on your work that you could include, such as evaluations from field supervisors or letters of recommendation?

Follow-through: Begin construction of your e-portfolio. Over the course of your career, you will be glad to have this archive of your work as you continue your learning and reflection process. You can also continue to add to your portfolio as your career unfolds and your learning and growth progress.

References

Baird, B. (1999). *The internship, practicum, and field placement handbook: A guide for the helping professions.* Upper Saddle River, NJ: Prentice-Hall.

Barrett, H. C. (2000). Create your own electronic portfolio: Using off-the-shelf software to showcase your own or student work. *Learning and Leading with Technology, 27,* 14–21.

Breault, R. (2000). Metacognition in portfolios in pre-service teacher education. Paper delivered to American Educational Research Association Conference, New Orleans, April, 2000 as cited in H. Barrett (2000). Create your own electronic portfolio: Using off-the-shelf software to showcase your own or student work. *Learning and Leading with Technology, 27,* 14–21.

Cantor, J. (1995). *Experiential learning in higher education: Linking classroom and community.* Washington, DC: ERIC Clearinghouse on Higher Education.

Chiaferi, R., & Griffin, M. (1997). *Developing fieldwork skills.* Pacific Grove, CA: Brooks/Cole.

Council for Standards in Human Service Education. (1995). *National standards for human service worker education and training programs.* Fitchburg, MA: Author.

Diambra, J., Cole-Zakrzewski, K., & Booher, J. (2004). A comparison of internship stage models: Evidence from intern experiences. *Journal of Experiential Education 27*(2), 191–212.

Dore, M., Epstein, B., & Herrerias, C. (1992). Evaluating students' micro practice field performance: Do universal learning objectives exist? *Journal of Social Work Education, 28,* 353–362.

Fry, R. (Ed.). (1988). *Internships: Advertising, marketing, public relations & sales.* Hawthorn, NJ: Career Press.

Gifford, A. P., & McMahan, G. A. (2001). Portfolio: Achieving your personal best. *The Delta Kappa Gamma Bulletin, 68*(1), 36–41.

Gladding, S. (1992). *Counseling: A comprehensive profession* (2nd ed.). New York: Macmillan.

Grant, R., & McCarty, B. (1990). Emotional stages in the music therapy internship. *Journal of Music Therapy, 27*(3), 102–118.

Grossman, B., Levine-Jordano, N., & Shearer, P. (1991). Working with students' emotional reaction in the field: An educational framework. *The Clinical Supervisor, 8,* 23–39.

Hartnell-Young, E., & Morriss, M. (1999). *Digital professional portfolios for change.* Arlington Heights: Skylight Professional Development.

Inkster, R., & Ross, R. (1998). Monitoring and supervising the internship. *NSEE Quarterly, 23*(4), 10–11, 23–26.

Kerson, T. (1994). Field instruction in social work settings: A framework for teaching. In T. Kerson (Ed.), *Field instruction in social work settings* (pp. 1–32). New York: Haworth Press.

Lamb, D., Barker, J., Jennings, M., & Yarris, E. (1982). Passages of an internship in professional psychology. *Professional Psychology, 13,* 661–669.

Michelsen, R. (1994). Social work practice with the elderly: A multifaceted placement experience. In T. Kerson (Ed.), *Field instruction in social work settings* (pp. 191–198). New York: Haworth Press.

Pettes, D. (1979). *Staff and student supervision.* Boston: George Allen & Unwin.

Siporin, M. (1982). The process of field instruction. In B. Sheafor & L. Jenkins (Eds.), *Quality field instruction in social work* (pp. 175–198). New York: Longman.

Sweitzer, H., & King, M. (2004). *The successful internship* (2nd ed.). Belmont, CA: Brooks/Cole.

Young, M. (2005). *Learning the art of helping: Building blocks and techniques* (3rd ed.). Upper Saddle River, NJ: Pearson.

Chapter 2

Getting Started

A tremendous amount of work must be done to get your internship underway. Even before your first day on the job, there is much for you to do. Simply selecting a fieldwork site can be time-consuming and complicated. If your program gives you input on your fieldwork site, there are many variables for you to consider in making this choice. The discussion and exercises in this chapter will engage you in this decision making by asking you to think about many of these variables carefully. The chapter also gives you the information you need to plan for and participate in the initial meeting with your supervisor. Finally, once your fieldwork has begun, one of your most important tasks is developing a learning plan. This chapter will acquaint you with the fieldwork learning plan and take you through the step-by-step process of developing one.

A Student's Reflections on Getting Started

It's hard to order when you don't know what's on the menu. That's how I feel when I try to decide where to do my internship. The human service field is so broad and varied, it is impossible to know about all the opportunities and careers that are out there. Where do I start? Sometimes I think that practically everything in this field interests me, so what does it really matter where I end up? Other times I get nervous about it and feel like this internship is setting the stage for the rest of my professional life.

continued

A Student's Reflections on Getting Started   *continued*

Whatever field I get experience with now will help me get a job in that field later. So I go back and forth between thinking that where I do my internship is the most important decision in the world and thinking that almost any place would be a good start.

Selecting a Field Site

Finding a field site that fits your learning needs is what selecting a field site is all about. Human service programs use a variety of methods for making student placements for fieldwork. In some cases, the student is entirely responsible for securing the placement, and the faculty member makes a contact with the agency to review the placement's appropriateness. In other cases, faculty members take responsibility for placing students, taking into consideration student interests, goals, and skills. In most cases, students have at least some input into their placement decision (Simon, 1999), but in some programs students may be assigned to placements with little to no input (Royse, Dhooper, & Rompf, 1996).

Students can learn a great deal even before the internship begins by being actively involved in the selection of a field site. Developing greater awareness of your career interests, building your job search skills, and enhancing your ability to communicate clearly and assertively are just a few obvious examples of learning that can take place in the field site selection process (Bertea, 1997).

Those programs that allow student input often ask students to submit information in writing as to the type of placement they would like to secure. It is not uncommon for student requests to be very general, for example, "I want to work with children," or unrealistically specific, for example, "I want to work in a residential treatment program with teenage substance abusers, and it needs to be within walking distance of the campus." These examples illustrate some of the most obvious considerations when thinking about a field site, such as the population served and the field of practice. You will be able to make a more informed decision about your field site by considering these as well as other factors more thoroughly.

A Student's Reflections on Choosing a Field Site

I have always wanted to work with children who have handicaps. I have said this for so many years, I think it might have just become habit. Now that I have a chance to test out some different careers, I am drawn to trying something different. I have worked at a camp for children with handicaps for the past few summers so I have some idea of what that type of work is like. While I'm a student and have this opportunity to do Practicum and Internship, I would like to broaden my horizons. I would be interested in trying out an administrative role of some kind. Also I think

continued

Considering Populations

Human service agencies work with people in need, providing services throughout the life span from birth to death. Due to specialization within the field, most agencies have a specific segment of the population as their target group. Many students identify the work that they wish to do in the human service field by identifying the age group with which they are most comfortable. This selection is often based on the student's previous job or volunteer experience, experiences with family members, or experiences in the student's own personal life.

Although all of these experiences form a good basis from which to consider various options, it is helpful to think more carefully about these options rather than choose a given population based on a general sense of comfort. As students are exposed to various populations, it is not uncommon for them to find that they enjoy working with a group that they had not considered before or perhaps had even avoided. Such experiences suggest that students would be wise to approach the process of field site selection with an open mind. The following exercise will help you to think through thoroughly and systematically the possibility of working with various populations. If you already have a field site, the exercise will help you to envision various scenarios that you might pursue in future internships, volunteer work, or employment.

◆ EXERCISE 2.1 PERSONAL REFLECTION: OBSERVATION OF SELF AND OTHERS

Consider each age group identified below and identify (1) the potential rewards of working with this group and (2) the potential frustrations. Briefly write down your thoughts and reactions as you respond to these questions. Although some groups will be initially more appealing to you than others, try to develop a balanced reaction to each population by responding to both (1) and (2) above.

Preschool Children (0–5 years)

Grade School Children (5–12 years)

Adolescents (13–18 years)

College-Age Young Adults (18–22 years)

Adults (23–65 years)

Older Adults (65–85 years)

Oldest Adults (>85 years)

Considering Fields of Practice (Settings)

Human service agencies vary broadly in their missions. An important step in selecting your field site is identifying the human problems or needs that you would like to learn more about. This dimension of an agency's or a professional's identity is referred to as the *field of practice* or *setting* (Barker, 1996). The field of practice shapes a worker's experience considerably as specialized expertise is required in each field. Given the range of human experience, it is impossible to identify all of the fields of practice here. The following exercise lists several major fields of practice for you to consider and react to as you develop your ideas about a potential field site.

◆ **EXERCISE 2.2 PERSONAL REFLECTION:**
   OBSERVATION OF SELF AND OTHERS

Think about your reaction to each field of practice listed below and briefly record your thoughts. Try not to focus on whether you already have the knowledge and skills to work in this field. Remember that you are entering the setting to gain these. As you consider each field of practice, respond to the following: (1) What interests you or attracts you about this field of practice? (2) What concerns, fears, or reservations do you have about working within this field? (3) Record your overall motivation to learn more about this population, using the ratings of Very High, High, Moderate, Low, Very Low.

Poverty

Homelessness

Child Abuse and Neglect

Child Development

Substance Abuse/Addictions

Mental Illness

Developmental Disabilities/Mental Retardation

Physical Disabilities

Education

Health Care

Crisis Services

Domestic Violence

Concerns Related to Aging (e.g., social isolation, physical/mental decline, retirement)

Crime

Legal Assistance

Social Justice and Social Change

Therapeutic Recreation

It would be useful to discuss your responses to Exercise 2.1 and Exercise 2.2 with other students and to have your work reviewed by a person who is experienced in the human services field, perhaps one of your faculty members. Sometimes initial reactions to various fields are based on misinformation or stereotypes. Sharing your thoughts with others helps to identify misconceptions, biases, or false assumptions that might be present.

As you select a field site, you will probably want to examine most closely those fields above that you responded to with at least "moderate" motivation. For each of these fields, think at greater length about what aspects of that field capture your interest. In some cases, human service workers are drawn toward a particular field in an effort to deal with unresolved issues in their own personal lives. For your own well-being, as well as that of your clients, you should not select a field site that will take you too deeply into troublesome issues that may be present in your own life. In such a situation, it is often difficult to sort out your own thoughts and feelings from those of your clients and to maintain appropriate boundaries and objectivity in your helping relationships. If this is a field that you want to enter, you may still do so after you have resolved your own issues either through professional counseling or personal reflection and effort. If you are uncertain as to whether you have resolved an issue sufficiently to begin professional work in that area, discuss your concerns with a trusted faculty member or a professional counselor.

Considering Agency/Worker Function

Human service agencies perform a variety of roles and functions in relation to any given population and field of practice. Agency function has to do with the nature of the goals and interventions within the organization. The local community often has a number of agencies working in different ways to benefit a given population. Also, within larger human services agencies there are typically specialized departments performing a variety of tasks to benefit the target population. For example, within a given community there may be a mental health center that provides direct services to clients through its counseling services, while another agency, perhaps the mental health association, provides community education and prevention programs. There may be a department within one of these organizations that leads fund-raising efforts and carries out other administrative and leadership functions to expand services for the mentally ill within the community.

In 1969, the Southern Regional Education Board conducted a study of human services workers that identified 13 different functional roles of human services workers (Mandell & Schram, 1985; Mehr, 1998; Neukrug, 2004; Woodside & McClam, 2006). Some agencies conduct many of these functions, while others may conduct only one or two of them. Considering each functional role listed below will help you to identify more clearly the types of experiences and skills you are seeking through your fieldwork.

 ◆ **EXERCISE 2.3 PERSONAL REFLECTION: OBSERVATION OF SELF AND OTHERS**

As you consider each of the following functional roles, write a response in which you identify: (1) your level of interest in this role, (2) your current level of skill in performing this role, and (3) how important you think it is to develop further skills in this role prior to completing your human service program.

Outreach—going into the community to work with clients in their environment, especially for the purpose of reaching underserved populations or for prevention

Broker—helping clients locate and access services

Advocate—assertively seeking services for clients and defending clients' rights; assertively seeking social change

Mobilizer—organizing community groups to develop needed services and/or initiate social change

Behavior change—using intervention strategies and counseling skills to facilitate client change

Teacher/educator—informing individuals and groups

Caregiver—offering direct care to clients physically, emotionally, and/or socially

Evaluator—assessing human services programs for effectiveness and financial accountability

Assistant to specialist—working closely with a highly trained professional as an aide in delivering services

Data manager—gathering information and organizing it for use in a variety of tasks involving direct services to clients and/or administrative services

Administrator—supervising human services agencies or programs

Community planner—identifying community needs and designing programs to meet those needs

Consultant—offering knowledge and expertise to other professionals and/or to client and community groups

---

Reflecting upon these functional roles helps you to identify the specific roles you would like to enhance in your own professional development. Although students usually begin thinking about their fieldwork by identifying the population or field of service they would like, it may be more useful in some cases to think first in terms of functional roles. These roles are generalizable from one population and field of practice to another. For example, advocacy skills learned through work with welfare clients can be easily transferred to working with abused children, the mentally ill, or any other client group. Such skills form the core of your development as a professional and, therefore, may well be of greater significance and lasting value than gaining experience with a given population or field of practice.

For more information about the roles and skills of human services professionals visit the websites for the Human Services Career Network and the National Organization for Human Services. There you will find further information about roles and skills of human service professionals.

◆ **EXERCISE 2.4 PERSONAL REFLECTION:
OBSERVATION OF SELF AND OTHERS**

Review the writing you have done in Exercises 2.1, 2.2, and 2.3. Keeping in mind your reactions to the various populations, fields of practice, and functional roles, describe at least three field placements that would be beneficial for you at the present time. Avoid simply naming agencies or organizations that you might know about. Instead, describe the type of learning experiences that you need. Draw upon the concepts of population, field of practice, and functional roles in your descriptions.

Faculty members and more advanced students in your program can help you identify organizations that offer the types of experiences that you describe. Researching various placements through visiting their websites also gives you more insight into their mission and programs. Depending upon your school's policies regarding making placements, you might contact a few agencies regarding your interests. The work you have done in the exercises of this chapter will help you articulate your learning needs and give you some ideas about where to start in your search for a field site.

The Initial Meeting with Your Prospective Supervisor

Once you and the faculty of your program have secured a tentative site for your fieldwork, you should schedule a meeting with your potential field supervisor (Faiver, Eisengart, & Colonna, 1995; Morrisette, 1998; Thomlison et al., 1996). This meeting is in your best interest because it helps to ensure that there is a good match between what the agency has to offer and your interests. Also, most supervisors want to interview students prior to approving the placement officially.

It is generally the student's responsibility to initiate and schedule this meeting. Students should prepare for the interview by learning as much as possible about the organization ahead of time. Checking out agency websites and speaking with faculty and students who are familiar with the organization enables you to enter the interview with greater confidence and with more focused questions about the experience you might have there.

Typically, the supervisor will want to focus on such issues as why you are interested in this particular site, what your career goals are, what experiences (both classroom and direct practice) you have had that might be relevant to the internship, any previous field placements you might have had, and what you are hoping to gain from the field experience. Additionally, you should give thought to questions that you would like to ask during this meeting.

Students are generally most interested in what they will actually do in the internship—the tasks they will perform, the roles they will assume, and the responsibilities they will have. These aspects of the internship vary from one agency to another and from one supervisor to another. Even the time of year that you perform your internship can alter your responsibilities. For example, a homeless shelter tends to have more clients and therefore greater demands in cold weather as opposed to warm weather, increasing the likelihood that an intern would work extensively with clients. A student in the same internship at a different time of year might have more responsibilities that involve planning, organization, and administration while having less client contact. The only way to know specifically what tasks your internship entails is to ask your supervisor directly.

Students are also generally interested in the nature of the client population served within the agency, the range of services and programs offered, the types of experiences the agency will make available, and typical expectations of students. Thomlison et al. suggest a number of questions students might ask their field supervisors during the initial meeting, including the following:

> What are the kinds of things I will be doing and learning here? . . . What systems will I be working with? . . . Will I have more than one [field] instructor? . . . What is the largest single challenge facing your organization (e.g., agency or setting) that I should know about? (1996, p. 47)

In addition to such questions, you might also want to raise a number of general orientation questions, such as:

- ◆ What is a typical day like in this setting?
- ◆ What are the typical roles and tasks of interns?
- ◆ How should I dress for the internship?
- ◆ What hours will I be working?
- ◆ Where will I be located physically in the facility?
- ◆ What arrangements, if any, do I need to make for lunch or other meals?

No doubt you will have additional questions based upon your own particular interests and concerns.

The initial interview is similar to a job interview in many ways, but it is usually more spontaneous and informal (Wilson, 1980). The meeting is somewhat like a job interview, however, because both you and the supervisor are deciding if a placement in that particular setting is right for you. As in a job interview, you will want to make a good first impression by arriving on time, dressing appropriately, making eye contact, and shaking hands firmly. In the interview you will, of course, want to convey your eagerness to learn and your interest in the organization while also conveying something of your own abilities and relevant background. If you have had previous experience in human services, it would be a good idea to prepare a resume reflecting these experiences and bring a copy for the supervisor.

Based upon this interview, the placement generally moves toward formal approval. If either you or the supervisor has serious reservations about the placement following this discussion, your faculty liaison should be notified as soon as possible. The concerns raised by the interview may be resolvable, but if not, an alternative placement will need to be arranged. To allow sufficient time for this possibility, it is important to schedule your initial meeting with the field supervisor in a timely manner.

A Student's Reflections on Meeting His Supervisor

I thought I was going to meet my supervisor just so we could start getting to know each other. Our conversation was more like a job interview than I expected it to be. For example, she asked about my educational background and practical experience, and she asked what I see as my strengths and weaknesses. Although it made me a little nervous, I see it as good practice for my job interviews later on. I am excited about the possibilities we discussed for my internship. It sounds like my supervisor is going to trust me and give me some real responsibility. After a few weeks I will even be able to have my own cases to work with independently. I do, however, have some concerns. The agency, and my supervisor in particular, seem to be very busy. I'm concerned that they might not have a lot of time for a student. I have given it some thought and have decided that I do want the placement nevertheless. I have no doubt that I can learn a great deal there, even if people are very busy. I learned so much just by going for a visit. Another smaller concern that I have is about how to dress. My supervisor was really dressed up, and I noticed that several other workers were as well. Most of the male staff members were wearing ties, and the women were wearing dresses and heels. Am I expected to dress like that? I meant to ask at the time, but it slipped my mind with all the other things to discuss. I plan to call her before my first day to ask about this.

◆ EXERCISE 2.5 PERSONAL REFLECTION: OBSERVATION OF SELF AND OTHERS

In preparation for the initial meeting with your field supervisor, write in the space below the questions you would like to ask or topics you wish to discuss in this meeting. Also include any information about yourself and your experiences that you would like to share with your supervisor.

An initial meeting with your supervisor in which your questions are answered will help you to arrive on the first day of the internship with greater confidence and comfort. It is well worth the time and effort involved as it helps you make a smooth transition into your new environment. Following this initial meeting, it is a good idea to follow up with a thank you note or letter to your field supervisor (Faiver et al., 1995). The note can either indicate your interest and eagerness to begin the internship or can convey that you have decided to pursue a different placement more consistent with your goals. In either case, your conversation with this human services professional was helpful to you in making a decision about your future and is worthy of a word of thanks.

Developing a Learning Plan

The initial meeting with your supervisor begins to lay the groundwork for your fieldwork learning plan. The learning plan is a written document in which your learning goals are specifically identified as are the strategies or activities for accomplishing each goal (Bogo & Vayda, 1998; Thomlison et al., 1996; Wilson, 1980). The learning plan also identifies how your progress toward each goal will be measured. This plan may be referred to within some human service programs as a *learning contract* or *learning agreement*. Whatever term is used, this document is an essential tool for guiding your fieldwork.

In field experiences that are a full-academic year in duration, the program may allow as much as a month to develop the plan. In experiences that are a semester or less, the plan must be developed more quickly, usually within one to two weeks. Once developed, the learning plan serves as an important road map, directing and organizing your experiences. Therefore, it is important that you develop your learning plan as early as possible.

In most programs, the learning plan is open to revision throughout the field experience. Contingent upon the mutual agreement of the student, field supervisor, and supervising faculty member, goals and strategies can be modified, added, and/or deleted within reason. As the placement progresses, new opportunities often arise and new interests develop. By the same token, there are also times when previously anticipated opportunities do not develop. While the integrity of the learning plan needs to be preserved in order for it to serve its purpose, it must also be a living document, capable of responding to changing circumstances.

A Student's Reflections on the Learning Plan

At first I looked at writing up a learning plan as busy work. Now that it's done, I definitely see the value of it. It helps me keep in mind what I'm supposed to be getting out of this. I think without the learning plan I wouldn't be nearly as clear about why I'm here, what I'm supposed to be learning, and what I'm supposed to be doing. I think it has also helped the staff see me as a student and not a volunteer. They take me more seriously, I think, because it's clear that I'm here to learn.

Identifying Goals

The development of the learning plan is a mutual process involving you and your field supervisor. Forming the basis of the plan is the discussion in your initial meeting with your supervisor in which you explained something about your interests and the field supervisor described the experiences that the agency could offer. You and your field supervisor now need to discuss in more specific terms the exact nature of your learning goals. Although it is your responsibility to identify your own learning needs and to generate your own ideas about your specific goals, your field supervisor's involvement is needed to ensure that these goals are realistic for the particular setting and for the time available. Additionally, as an expert in the field, your field supervisor can ensure that you do not omit important areas of learning.

The level of student responsibility for shaping the fieldwork learning experience often requires a major adjustment for students. In a traditional course, the professor identifies course objectives, makes assignments, and informs students as to how they will be evaluated. Developing the learning plan requires that students take far more leadership in shaping their learning process. In creating the learning plan, you will initially produce a draft of your goals, the strategies that you will use to accomplish them, and the methods that will be used to measure your progress. Your field supervisor can then review this draft and make suggestions to further shape and develop the plan. You should take the time to develop your learning plan carefully, not only because it will direct your learning in the internship but also because it can be used for legal purposes in some circumstances (Gross, 1993).

To develop your learning plan, you might begin by reading the general goals for the internship that have been developed by your human services program. These goals are generally written in either the syllabus for the internship or in a student handbook, or both. The program's goals for the internship will help you to identify the broad parameters within which your learning should operate. Some programs direct students to set goals in specified areas. Structuring the goal-setting process in this way is helpful because it guides students to develop a well-balanced plan for their learning. For the purposes of the exercises in this chapter, you will develop goals related to the following three categories of learning:

1. **Knowledge:** defined as learning and understanding facts, concepts, theories, information, or ideas.
2. **Skills:** defined as the ability, gained through practice, to perform tasks effectively.
3. **Personal development:** defined as attitudes, values, biases, and/or habits that are uniquely your own.

Goals in each of these three areas should be as specific as possible.

Identifying Strategies

After setting your goals, you must then identify specific strategies that you will employ to accomplish each goal. To understand the difference between goals and strategies, keep in mind that a goal is an outcome or an endpoint, not an activity, program, or event. To state each goal as an outcome, you must ask yourself exactly what knowledge, skills, or personal development you hope to have attained by the end of the internship. For example, "I will become familiar with the statutes in North Carolina pertaining to domestic violence" is a good goal statement. "I will attend a conference on domestic violence" is *not* a goal statement but is a good statement of a strategy. As with your goals, your strategies will be most useful if they are as specific as possible. Strategies might include such activities as observing, attending meetings, working with clients, interviewing co-workers, reading, and so on.

Identifying Methods of Assessment

As the final step in developing your learning plan, you and your field instructor will specify ways to measure your progress. Ideally, the ways you select to measure your progress will rely upon information that is directly observable by your field supervisor and/or your faculty liaison because they share responsibility for evaluating your work. The methods you use may vary with each goal. For some goals you might use only one method, whereas for other goals you might use several.

Assessment methods often include such activities as direct observation of student work, supervisor review of written documents that the student has produced, supervisor review of student videotapes or audiotapes, and/or supervisor or faculty liaison observation of student presentations. Internship students often write papers pertaining to their work and/or keep daily journals of their activities, observations, and reflections. These documents, too, might be reviewed and evaluated by the supervisor and/or faculty liaison as a method of assessment for particular goals. As you construct your learning plan, your field supervisor or faculty liaison might have additional suggestions regarding how to evaluate your progress on the various goals that you set.

As with the other elements of your learning plan, collaboration with your field supervisor and/or faculty liaison may lead to some revisions in your methods of measurement.

Nonetheless, you should develop and present your own ideas about this important aspect of your learning plan. This might seem odd to you because throughout your education faculty members have determined how your learning would be assessed and measured. As an internship student who is making the transition into professionalism, you can appropriately join this conversation by developing your own ideas about methods of assessment. This is a significant step in taking responsibility for your own learning.

Below is a sample learning plan that includes examples of goals, strategies, and methods of measurement. Although this sample lists only one or two goals under each subheading for the purposes of illustration, an actual learning plan generally includes several goals under each subheading.

Learning Plan Example

Student: Marianne Harris   **Field Supervisor:** Jean Miles
Agency: Turning Point Domestic Violence Services

Knowledge Goals	Strategies	Methods of Measurement
Goal 1: Learn about state statutes regarding domestic violence	a. Attend conference on this topic b. Observe staff members as they explain legal rights to clients c. Attend court and observe proceedings d. Read *Your Legal Rights: A Guide for Survivors*	a. Write newsletter article about this topic b. Discuss what I've learned with field supervisor c. Prepare PowerPoint presentation about legal rights of abused women for presentation to Board of Directors
Goal 2: Learn about the impact of domestic violence on children and the needs of these children	a. Work with children in the shelter b. Work in the crisis nursery c. Read scholarly journal articles on this topic	a. Will turn in summaries and critiques of at least three relevant journal articles b. Will plan and lead at least one activity with the children in the shelter and get feedback from staff c. Will plan and lead at least one activity with children in crisis nursery and receive feedback from staff

Skills Goals	Strategies	Methods of Measurement
Goal 1: Improve my ability to implement crisis intervention strategies	a. Observe my supervisor and other staff members working directly with crisis victims b. Go through the crisis intervention training program that is offered for volunteers c. Staff the crisis line and work directly with at least three clients who are in crisis	a. Discuss in supervision my observations of staff skills and behaviors b. Write summaries of my interactions with clients and discuss them in supervision

Personal Development Goals	Strategies	Methods of Measurement
Goal 1: Become more assertive in working relationships	a. Interview workers about how they handle difficult situations with clients and co-workers b. Read about assertive communication and professionalism c. Practice assertiveness skills daily in both my personal and professional life	a. Keep a journal in which I reflect on each day's interactions and evaluate my own assertiveness b. Write at least one paper in which I examine my assertiveness using the Integrative Processing Model

EXERCISE 2.6 ANALYSIS

In this exercise, you will begin developing your personal learning plan for the internship. For the purposes of this exercise, identify below at least two of your learning goals for each of the three categories discussed earlier: knowledge, skills, and personal growth. For each goal, list the strategies that you will use to achieve that goal. A well-developed learning plan identifies several goals for each category of learning (knowledge, skills, and personal development) as well as strategies for achieving each goal and methods of measurement.

Learning Plan

Student: _____   Field Supervisor: _____

Agency: _____   Date: _____

Knowledge Goals	Strategies	Methods of Measurement

Skill Goals	Strategies	Methods of Measurement

Personal Development Goals	Strategies	Methods of Measurement

As you continue to develop your learning plan, use the process and steps that you have learned in this chapter. When you have completed a draft of your learning plan that represents your best and most complete effort, make a copy of it for your field instructor and schedule a time to discuss it. This conversation will probably produce modifications and revisions, leading to your final draft. Both you and your field supervisor should sign the learning plan in its final form to indicate your mutual support for the plan and your intention to implement it.

Once your learning plan is complete, you will find that you feel more focused in your work and that you can better articulate what you are learning from your fieldwork. As you implement your learning plan, discuss with your supervisor how well the plan is serving your learning needs. If you feel that the plan needs modification as you go along, express this to your supervisor as well. The learning plan should always enhance your learning by providing focus, direction, and structure.

Conclusion

Beginnings are important. This chapter provides much of the information you need to launch your internship successfully. Considering carefully your learning needs and interests as you select a potential field site, exploring important issues in your initial meeting with your field supervisor, and developing a well-thought-out learning plan are essential steps in laying a solid foundation for your internship.

◆ FOR YOUR E-PORTFOLIO

What are your feelings and concerns as you begin your internship? What do you see as your primary learning interests and needs? What are the issues that are most important to you as you plan your internship experience with your supervisor? What aspects of your learning plan are you most excited about? What aspects do you believe will be most challenging?

Once your learning plan is developed, add it to your e-portfolio. Refer to it often, tracking your progress toward your identified goals.

References

Barker, R. (1996). *The social work dictionary*. Washington, DC: NASW Press.

Bertera, E. (1997). Empowering human services students to select their practica: Insuring goodness of fit. *Human Services Education, 17*(1), 65–67.

Bogo, M., & Vayda, E. (1998). *The practice of field instruction in social work: Theory and process* (2nd ed.). New York: Columbia University Press.

Faiver, C., Eisengart, S., & Colonna, R. (1995). *The counselor intern's handbook*. Pacific Grove, CA: Brooks/Cole.

Gross, L. (1993). *The internship experience* (2nd ed.). Prospect Heights, IL: Waveland.

Mandell, B., & Schram, B. (1985). *Human services: Introduction and interventions*. New York: Macmillan.

Mehr, J. (1998). *Human services: Concepts and interventions strategies*. Boston: Allyn & Bacon.

Morrisette, P. (1998). The undergraduate preinternship process: Issues and recommendations. *Human Service Education, 18*(1), 49–55.

Neukrug, E. (2004). *Theory, practice, and trends in human services* (3rd ed.). Pacific Grove, CA: Brooks/Cole.

Royse, D., Dhooper, S., & Rompf, E. (1996). *Field instruction: A guide for social work students* (2nd ed.). White Plains, NY: Longman.

Simon, E. (1999). Field practicum: Standards, criteria, supervision, and evaluation. In H. Harris & D. Maloney (Eds.), *Human services: Contemporary issues and trends* (pp. 79–96). Boston: Allyn & Bacon.

Thomlison, B., Rogers, G., Collins, D., & Grinnell, R. (1996). *The social work practicum: An access guide*. Itasca, IL: Peacock.

Wilson, S. (1981). *Field instruction: Techniques for supervisors*. New York: Free Press.

Woodside, M., & McClam, T. (2006). *An introduction to human services* (5th ed.). Belmont, CA: Thomson.

Chapter 3

Getting Acquainted

As a newcomer, you have an ideal opportunity to gather information about your field-work organization. You will be introduced to many staff members, most of them eager to tell you about the agency and to answer your questions. You will probably be given a general orientation to the agency and offered printed materials to peruse. Also, in the first few days of your internship, you will likely have some unstructured, unscheduled time that you can use to get acquainted with the organization and the people within it.

These common rituals and rhythms of getting started open the door for your exploring and learning as much as you can about your field site. In some circumstances, the nature of the work may be such that you do not have the luxury of a slow startup or a thorough orientation. Then, you need to rely on your own observations, assertiveness, and resourcefulness to get oriented. In either case, this chapter helps you to focus and direct your attention toward the specific aspects of agency identity that make each organization unique and interesting.

A Student's Reflections on Getting Acquainted

My entry into the internship has been one of mixed emotions. At times, I've felt alone, as though I am a stranger looking in. The counselors don't really have time to play "nice to meet you." The residents come first, and I realize that this is how it should be.

continued

A Student's Reflections on Getting Acquainted   *continued*

Nevertheless, I have jumped in with both feet and am learning a lot about the place as I go along. My endless questions are always answered, and no one is afraid to give me directions or instructions. Overall, I have been welcomed with open arms. But like any new job, it takes a while to learn the ropes. I have to realize that I can't learn everything in one day. I'm glad to be here. Every morning I am faced with a challenge that I look forward to conquering.

Getting to Know Your Agency

Getting to know as much as you can about your agency early on not only helps you feel more comfortable and confident in your new setting, it also helps you to understand and integrate your future experiences as they unfold during your internship. Some of the most useful and accessible means of gathering information about your field site in the earliest days of your placement are to

- ◆ research the agency's website.
- ◆ read pamphlets, brochures, or any other written information for distribution to the community.
- ◆ look over the agency policy manual.
- ◆ read any periodic reports that the agency prepares for its governing and/or funding bodies—for example, budget proposals, annual reports, grant proposals, and so on.
- ◆ formally interview or informally speak with key people in the organization, such as the agency director, the human resources director, department heads, and employees with especially long seniority.
- ◆ attend any of the routine meetings of the organization that you are allowed to attend.

Discuss with your field supervisor your interest in learning more about the agency and determine which of these sources of information you can access in your organization. Your field supervisor might also have additional resources to suggest. Of course, one of the best ways to learn about your organization is by simply being a good observer. Keeping your eyes and ears open and your mind engaged at all times is your best information gathering tool.

◆ **EXERCISE 3.1 ANALYSIS**

Describe your first impressions of your field site. Describe its physical appearance both inside and out as you made your first visit. What messages and values seem to be conveyed to clients through these physical characteristics? How were you met and greeted by the staff in the organization? What were your observations about how clients are met and greeted? What messages and values are conveyed through these behaviors?

Getting to Know Your Co-Workers

In your first few days in the organization, you may be overwhelmed with new faces and new names, especially if your organization is large. Allow yourself some time to simply observe the various people working around you and their interactions. It is, of course, understandable that you might be nervous at first about approaching your co-workers for conversation. Try to identify those workers who may be especially approachable and take a few risks in extending a hand and introducing yourself. You will, of course, want to approach those workers whose job, education, or professional background particularly interest you in some way. Gathering the information for the exercises in this chapter will give you a good focus and an incentive for approaching your co-workers. Most professionals are pleased, even flattered, to find that a student is interested in their work. Nevertheless, in approaching staff members, be respectful of the fact that they may not have time for a lengthy conversation at that particular time. It may be best to ask for a scheduled time to meet if you find that you have several questions to discuss.

A Student's Reflections on Getting Acquainted

Every day I learn more and more about this place. There is so much more here than meets the eye at first. I never knew, for example, that this organization was started by the Lutheran Church (my own denomination). Also I recently met an employee

continued

Getting Acquainted

43

who is running for public office because she wants to do more to influence social policy and human service funding in the state. She has opened my eyes to how government policies on different things affect the quality of life. I have never been interested before in government policies on issues like minimum wage, childcare, and health insurance. This internship is really helping me to see and understand the "big picture."

As a general rule, you enhance your experience in your internship if you try to stay in the flow of interaction within the organization, getting to know as many people as possible. If the staff has a gathering place for lunch and breaks, try to be there for those times as much as possible. If you have your own office, keep the door open rather than closed whenever you can. If your supervisor tells you that you may attend a certain meeting "if you would like," attend it if at all possible, even if it is at an inconvenient hour. Make yourself as approachable and involved as possible, remembering that staff members, too, may have some anxieties and uncertainties about approaching you.

As you make connections with your co-workers, you are engaging in the important, but generally untaught, career skill called *networking*. Networking is the art of using your interpersonal skills to forge links with others. Networking may be defined as "the relationships professionals cultivate with other professionals to expedite action through the social system" (Barker, 1996, p. 253). In human service organizations, like all organizations, a great deal occurs through informal channels. Having good relationships with colleagues helps you to provide better services and linkages for clients as they move through the human services system. These relationships can also provide excellent sources of informal mentoring, support, and guidance for your own professional development and future job search (Neukrug, 2004).

In getting to know your co-workers, do not overlook support staff. Secretaries and administrative assistants often have an excellent overview of the organization and know a great deal about interacting with clients based on years of experience. Maintaining a positive relationship with support staff also enables you to do your job better due to the support and information that they can provide (Thomlison et al., 1996).

◆ EXERCISE 3.2 ANALYSIS

Identify at least one staff member whom you have met and talked with in the first days of your fieldwork. Identify the individual by name and job title, summarize your conversation, your impressions of the individual, and your reaction to the conversation. Also identify particular people in the organization whom you have not met but plan to meet within the next few days or weeks. In identifying these individuals, think of gaps in your current knowledge and understanding of the agency. What perspectives are you missing?

Understanding Your Role in the Organization

The process of clarifying your role within the organization most likely began even before the internship began. When you interviewed with your supervisor prior to the internship and discussed your placement with your faculty liaison, general ideas about your role and responsibilities as an intern were probably discussed. As you learn more about your agency and your co-workers, your own role within the organization will become clearer.

Nevertheless, interns can often suffer from some degree of role ambiguity. Because most interns have previously held jobs and understand what it means to have a job, they often expect the internship to be like a job. In reality, most internships are not very similar to a job. When an organization hires a staff member, there is generally a clear job description that identifies a set of responsibilities for that employee. The agency in this case has a clear set of needs that the employee is expected to fulfill. Some internships are developed in this way, with a clear job description for the intern.

Most internships, however, are much more flexible and perhaps even somewhat vague in their expectations of students. Although this can be a bit unsettling for students who would like more structure and role definition, there also can be great learning advantages to this approach. Internships of this type can work very well as long as all concerned remember that the intern is a student who is there primarily to learn. A more flexible internship of this type might allow you to move from one department to another and from one worker to another in order to be involved in the most valuable learning experiences as they emerge. Many times this flexible approach is coupled with a major project that the intern is assigned with the expectation of a "product" at the end of the term. In other internships, the intern's time may be primarily devoted to one task, department, service, or worker with opportunities to venture into other areas as appropriate.

Unfortunately, in some cases, there are times when interns enter organizations that are overwhelmed with the demands of daily operation, and as a result, the intern's role is not given sufficient clarity and focus. If this is the case two to three weeks into the internship, you should discuss your concerns with your field supervisor and faculty liaison. Often faculty members have a history of working with the organization and can provide information to help you place your own experience within the larger context of the "typical" student experience. Faculty liaisons generally want students to handle issues as independently as possible, but when it is appropriate, they can also advocate for students in order to ensure that their learning needs are met.

◆ EXERCISE 3.3 PERSONAL REFLECTION: OBSERVATION OF SELF AND OTHERS

Describe your role within the organization to the best of your ability. How satisfied are you with the role that has emerged for you thus far? What issues, if any, would you like to discuss with your field supervisor in order to clarify or modify your role?

Learning About the Agency's Mission, Goals, Objectives, and Strategies

A clear statement of an organization's mission, objectives, goals, and strategies is central to its efficient functioning. Your understanding of your field site is enhanced by examining these elements of its identity. These elements may be predetermined through legislative or funding mandates or may be determined more locally by the staff, administrators, and/or board of directors.

Human service agencies often have a formal, written mission statement in the first pages of their policy manual. The prominence of this statement reflects its importance to the agency's identity. The mission statement of an agency sets out the rationale for its existence, explaining whom the agency is intending to serve and the community needs or problems that it seeks to address. The mission statement also conveys important information, directly or indirectly, about the agency's philosophy of helping and the values that underpin its efforts.

Closely related to the agency's mission are its goals and objectives, as well as the strategies it uses to meet them. The agency's goals can often be found in close proximity to its mission statement. The goals of the agency describe the outcomes that it is trying to accomplish, often using broad, difficult-to-measure terminology. The goals are effective in giving the agency direction toward the long-term achievement of its purpose. In contrast, objectives are written or discussed in more specific, measurable language and are effective in focusing and directing more immediate efforts. For example, an agency may have the goal of assisting the homeless in the community and the objective of reducing the number of homeless people in the community by 10 percent by the year 2012 (Barker, 1996). Many agencies have written goals and objectives for the organization as a whole as well as for each department of the agency, while other agencies are less formal and comprehensive in their administrative methods.

Agencies employ specific strategies to fulfill their missions and accomplish their goals and objectives. Programs, services, and activities are planned, designed, and implemented toward this end (Weiner, 1990). The strategies include not only the services and programs directed toward clients but also such behind-the-scenes administrative efforts as supervision, fund raising, staff training, and continuing education.

Collectively, the mission, goals, objectives, and strategies of the organization reflect the organization's philosophy and values. In some cases, this philosophy is explicitly stated in the mission statement, while in others it is more implicitly suggested. Often, it is the agency's strategies that are most telling in this regard. Several agencies might serve identical populations and seek identical outcomes but have vastly different ideas and philosophies about how to reach those goals. Consequently, their strategies will be vastly different as well.

For example, in the case of three programs serving teens with substance-abuse problems, each might pursue its goal of helping clients become drug free by different approaches. The first might encourage the youth toward spiritual transformation and reliance on a Higher Power. The second might use traditional psychotherapy in combination with carefully administered psychotropic medications. The third might use outdoor adventure experiences, hard work, and group cooperation. Whether or not their philosophies are spelled out, clearly there are different ideas and values operating in each agency's understanding of the nature of the individual's problem as well as the strategies that are thought to be most effective in prompting change.

The philosophy and values of an organization are compellingly powerful determinants of *how* the agency goes about accomplishing its goals. As you examine your agency's mission statement, identify the philosophy and values that guide the agency's work and shape its identity.

A Student's Reflections on the Agency's Mission Statement

As you can imagine, I was less than thrilled when my supervisor suggested that I spend the first hour or two of the day just "getting familiar with our printed material." I cast my gaze upon a book, three inches thick, in a drab gray three-ring binder. Some of it was dry policies and such, but the agency's mission statement really

continued

<table>
<tr><td>

A Student's Reflections on the Agency's Mission Statement *continued*

inspired me. I have since found myself thinking about that statement, particularly when I feel discouraged about my work. In the daily grind it's easy to lose sight of what we're really trying to accomplish here. I think if I were an employee here with my own office and desk, I would frame a copy of the mission statement and keep it at exactly eye level right over my desk.

</td></tr>
</table>

 ◆ **EXERCISE 3.4 SYNTHESIS: LINKING KNOWLEDGE AND EXPERIENCE**

Explain, in your own words, the purpose or mission of your agency.

Describe the goals and objectives of your agency or, if you are in a very large organization, those of your department. If your agency's goals and objectives are not in written form, interview a few key people in your agency and inquire about the goals and objectives that are informally guiding the agency's efforts.

As you observe the agency's work, what specific strategies are being used to meet the agency's goals and objectives?

What do the agency's mission statement, goals, objectives, and/or strategies say or imply about its philosophy and/or values? How are the philosophy and values reflected in the agency's operations?

Based upon your observations thus far, what is this agency actually accomplishing for individuals? Families? The community?

Learning About the Agency's Organizational Structure

Your agency, unless it is very small or very new, probably has an organizational chart that identifies by job title the various positions in the agency and reflects who reports to whom within the system. The organizational chart conveys information about how the agency is organized to accomplish its work. Divisions, departments, and programs within the agency appear on the chart in such a way that the various units of the organization and how they relate to one another become clearer.

Larger and more traditional organizations tend to follow a highly bureaucratic organizational structure, resulting in a chart that resembles a pyramid. The bureaucratic pyramid structure results in a vertical organizational chart, showing one person at the top, many people at the bottom, and an assortment of midlevel managers, coordinators, program directors, and supervisors in between. Smaller organizations and those with less traditional nonhierarchical management philosophies often have organizational charts that are more horizontal than vertical. These organizations use more peer supervision and collaborative methods of organizing and conducting their work.

An equally important factor to consider in organizational structure is the nature of the formal organization as compared to the informal organization (Alle-Corliss & Alle-Corliss, 1998; Ehlers, Austin, & Prothero, 1976; Lauffer, 1984; Russo, 1993; Sweitzer & King, 2004). Although the agency has an official organizational structure that reflects who reports to whom, who works with whom, and who has power relative to whom, the actual working of the organization does not always conform to this official structure. The patterns of communication, relationships, and influence that emerge in the day-to-day work of the agency are referred to as its *informal* organizational structure. The informal structure of the agency consists of a set of unspoken rules relating to every aspect of agency life, including how staff members relate to one another, how they relate to clients, how decisions are made, and so on. In general, organizations tend to function most effectively when the formal and informal organizational structures closely correspond to one another (Ehlers et al., 1976; Halley, Kopp, & Austin, 1998; Scott & Lynton, 1952).

As you become acclimated to the environment of your agency, you would be wise to observe both its formal and informal operations. Your ability to "fit in" with a particular agency depends in part on your ability to discern and follow the currents and crosscurrents of its formal and informal operations. The situation can at times leave a student feeling confused as to when to do what: Is it appropriate to talk to someone other than my supervisor about a given concern? Should I dress differently on days of particular events or meetings? Should I answer that phone or let it ring?

By being a careful observer of the informal operations and interactions within your agency, you will probably be able to figure out how to conduct yourself in most situations. Another good rule of thumb is "when in doubt, ask." Your supervisor is the most obvious resource but certainly not the only one. Draw upon the relationships you are developing in the agency and raise questions as they arise. (Similar issues regarding agency functioning will be further discussed as aspects of agency culture in Chapter 7, Dealing with Diversity.)

 ◆ **EXERCISE 3.5 ANALYSIS**

Secure a copy of the organizational chart for your organization. If the agency does not have a current organizational chart, try your hand at developing one and have your supervisor critique it. Describe in your own words the formal organizational structure of your organization. Explain how this structure serves to help the agency accomplish its purpose.

Describe the informal structure of your agency, based upon your observation. How do the formal and informal organizational structures compare with one another? How is the agency's work affected by its informal structure?

Learning About Agency Funding

Most human service agencies today are funded by a patchwork of sources. Both public and private agencies generally receive revenues from multiple coffers. Depending upon its funding sources and its mission, your organization might be considered (1) a public agency, (2) a private nonprofit agency, or (3) a private for-profit agency.

Public agencies are funded primarily by tax revenues at the federal, state, and/or local level. Although these organizations may receive private donations and user fees, they do not rely heavily on these revenue sources. These programs exist due to legislative mandate and include such agencies as departments of social services, mental health centers, public health centers, and public schools, among others.

Private nonprofit organizations are funded primarily by private sources, such as religious organizations, foundations, individual contributions, grants, and fund-raising proceeds. These organizations may receive some federal, state, or local revenues as grants for specific initiatives but not usually for ongoing operating expenses. In some cases, the government purchases services from private agencies. In such situations, the government establishes clear and specific contracts with these agencies to provide certain services to the public for which the government will pay (Ortiz, 1995). Many private nonprofit organizations have benefited from such government contracts for several years. In recent years, government contracts have become more common with private for-profit agencies as well (Gibelman, 1995). Also, depending upon the nature of the services provided, some agencies may qualify for reimbursement for their services from third-party payers, such as medical insurance companies and Medicaid, Medicare, or other government programs. Private nonprofits often receive grant money from private foundations and/or government sources as well. Some nonprofits may also charge small user-fees for specific services. Private nonprofits vary in size and complexity and include, for example, retirement centers that are funded by large religious organizations, large national organizations such as the American Red Cross, and smaller community organizations such as homeless shelters or crisis intervention agencies (Barker, 1996; Dobelstein, 1978).

Private for-profit organizations, sometimes referred to as *proprietary* social agencies, are initially funded by one or more individuals who invest their personal resources into the development of the organization. Examples of proprietary social agencies include private inpatient mental health and substance abuse facilities, educational institutions, nursing homes, camps, and so on (Barker, 1996). Once established, these organizations pursue a dual mission, to deliver services to a given population and to make a profit for their investors. Like some private nonprofit agencies, some for-profit organizations may also receive reimbursement for their services from state, federal, and/or local funds. Common examples are proprietary nursing homes that receive Medicare and Medicaid reimbursement (Garner, 1995) and for-profit day-care centers that receive reimbursement from JOBS or "Workfare" programs for providing care for the children of their clients (Hagen, 1995). Additionally, some of these organizations also may qualify for third-party reimbursement from medical insurance companies. For-profit organizations also often have some clients who pay "out-of-pocket" for services. In general, proprietary organizations charge higher user fees than do public and private nonprofit organizations.

Depending on the type of agency in which you are working, the administration may be willing to provide a great deal of detail about its funding and budget or may be willing to discuss this only in more general terms. Agencies (both public and private) that receive government funds are required to be more open with this information due to issues of public accountability.

An agency's funding method has far-reaching implications for its operations. Some agencies must put a great deal of time into fund raising, others have to work diligently to market their services, and still others must lobby and advocate for their budget needs through working with elected officials. It is difficult to make any generalizations about program quality based upon funding methods. Not-for-profit organizations are often seen as so underresourced that their services suffer. For-profit organizations are sometimes viewed suspiciously as the public questions the level of program quality that can emerge from a profit motive. Public agencies are often seen as being rigid, unresponsive, and overly bureaucratic. Each type of agency has its unique challenges, advantages, and disadvantages. It is my impression that there are excellent agencies within all three funding types and less effective agencies within all three types as well.

General information about public and nonprofit organizations can easily be accessed through websites. For example, the Chronicle of Philanthropy provides information relevant to nonprofit organizations, and the American Public Human Services Association provides information regarding public human services. General information regarding for-profit services might be more difficult to locate on the Web although websites do exist about particular companies that provide services of various types such as nursing care for the elderly, child care, mental health services, and others.

 ◆ **EXERCISE 3.6 ANALYSIS**

Classify your agency using the three categories described above. In as much detail as possible, explain how your agency is funded. How are the daily workings of your agency influenced by the nature of its funding? What advantages and disadvantages does the funding method create for your agency?

Learning About the Agency's Network

Because no one agency can serve all of the needs of its clients, human service agencies form an interdependent web of support and cooperation with one another (Halley et al., 1998). These relationships are generally informal, built from repeated interactions over long periods of time. At times, these relationships become especially cohesive and formalized as agencies pool their resources to accomplish some mutual goal, such as solving a community problem, providing a particular service or program, or even buying and sharing a piece of expensive equipment. Also, just as agencies cooperate to accomplish the work of serving the community, at times similar agencies may compete for resources, clients, or public recognition. Agencies serving the same target group may have difficulty establishing clear boundaries, delineating the identity of the programs as distinct from one another. Agencies may clash in their efforts to serve clients if those agencies have different philosophies of helping or different ways of viewing and conceptualizing client problems.

Much like the functioning of a family or any other social group, a certain amount of interagency conflict and competition is normal and predictable. Conflict and competition should not be seen, in and of themselves, as problems. As in other groups, tension sometimes leads to productive growth and change. Agencies that work well together over time find ways of identifying the sources of conflict and resolving them productively (Brueggemann, 2006).

A Student's Reflections on Agency Networks

I began the day by attending the CAP (Community Alternatives Program) Council on Aging breakfast with my supervisor. The council is comprised of members of approximately 20 different organizations. Each of the representatives reported on

continued

◆ **EXERCISE 3.7 ANALYSIS**

To answer the questions in this exercise, you will probably find it helpful to talk with various staff members in your agency.

Identify the particular agencies that work closely with your field site. When referrals are made by workers in your agency, where are these referrals most commonly made?

What agencies in the community refer clients to your agency?

Describe the nature of the relationship between your agency and the other agencies with which it interacts. In what ways do the agencies support one another? In what ways are conflict or competition present in these relationships, and how is this handled by the agencies involved?

How have the agency's relationships with other agencies developed and changed over time?

Learning About Your Client Groups

Most agencies have a particular client group that they serve. Your agency might deal with the poor, preschool children, troubled adolescents, or the recreational needs of healthy older adults. Many agencies will serve more than one client group. As part of your field experience, you will want to make every effort to learn more about the groups served by your agency. In some cases, you will have studied the relevant populations in your course work. In others, you might have had some exposure through your own personal life or that of a friend, family member, or acquaintance. In any case, you will want to be as knowledgeable as possible about the populations served.

You can refresh your knowledge by reading some current professional literature relevant to serving these client groups. All human services professionals and students are encouraged to read professional literature regularly as a method of expanding their knowledge about this and many other topics relevant to practice (Long & Doyle, 2004). Even if you have studied the population in question, you may have additional questions that you would like answered, or you might be curious about the latest research on particular issues or current methods used to address certain problems.

You might consider developing a reading list for yourself to answer the questions you have generated. Many internship faculty liaisons or field supervisors require their students to read professional journal articles that are pertinent to the field site's mission, services, or clients. Through your college or university's library you can locate numerous current scholarly journal articles and/or books that pertain to your internship. In doing your research, you should try to find publications that address some of the questions you have raised to make your reading more meaningful to you.

As your internship progresses and new questions are raised, you should continue self-directed reading designed to answer your questions. In fact, bringing relevant readings to your field site each day is a good habit that will ensure that you can use any unscheduled time productively. One of the pitfalls of formal education is that students can get in the habit of just reading what "an expert," such as a faculty member, has assigned. As a professional, you will have the responsibility to take charge of your own learning and professional growth. Your internship is the ideal time to start developing this habit through self-directed reading.

In addition to scholarly journal articles about your client population, websites can also offer valuable information. Particularly helpful are websites of research and advocacy groups for specific populations. Some of these may be government organizations while others are independent nonprofit groups. For example, the website for the National Institute for Mental Health, a federal government program, provides extensive information about current research pertaining to mental illness. Similarly, the Substance Abuse and Mental Health Services Administration is a governmental organization with an extensive website that includes information on far-ranging topics— from homelessness, to disaster relief, to schizophrenia, and HIV/AIDS.

Nonprofit advocacy groups also provide excellent information on the Internet. For example, the Alzheimer's Association is one of many organizations that represents the needs and interests of a particular population. Like many of these organizations, the Alzheimer's Association's website provides basic information about the disease as well as about research initiatives, resources, medical interventions, prevention, and more.

Finally, professional organizations' websites can be quite helpful. These organizations frequently provide information about the populations served by their professional members as well as information about relevant policy initiatives. For example, the National Association for the Education of Young Children provides useful information about the educational needs of young children as well as current research in the field and information regarding relevant public policy issues. Similarly, the American Public Health Association website includes up-to-date information regarding health concerns of various populations along with discussions about relevant legislative and policy issues.

 ◆ **EXERCISE 3.8 ANALYSIS**

Briefly identify and describe the client populations that your field agency serves. List 5 to 10 questions that you currently have about these populations, their problems, and/or the human services interventions currently used to serve them. Identify specific sources you might use to answer these questions.

As your internship and related research develop, some questions will be answered and new questions will be raised. Discuss your unanswered questions with your field supervisor, your faculty liaison, and your classmates. They may be able to direct you to additional resources to help fill the gaps in your knowledge. In some cases, it may be that your questions have not yet been answered in a systematic, scientific manner. In such situations, you may be in an excellent position as an intern to develop some tentative hypotheses and make preliminary observations on that topic. Practitioners with research skills can and do add significantly to our knowledge base in the human service field. Your fieldwork could provide a good opportunity to develop research interests that you might pursue in the future or even during your fieldwork with the approval and guidance of your field supervisor and faculty liaison.

Learning About the Context of Your Organization

Every agency exists within a larger context. The realities of the local community as well as the larger legal, social, and political environment at the state and national levels can have a significant impact on life within the organization (Kerson, 1994). Such issues are important because they set certain boundaries on what the organization can and cannot do, which, of course, affects what you can and cannot experience as an intern in that setting. The nature of the local community and legal, social, and political issues can profoundly influence the functioning of the agency as well as staff morale, stress, and workload. Through considering these topics, you will gain a better understanding of the supports, tensions, and stresses that come into play as your agency performs its work against a backdrop of various forces.

Getting to Know the Community

Your first step in understanding the community context of your organization is simply identifying the community that your agency serves. This task might not be quite as straightforward as it seems at first glance because the community context for the organization is not necessarily synonymous with the town or county in which it resides.

The agency's funding and mission are primary factors in drawing the parameters around the agency's community. For example, a state-funded school for children with behavioral problems may have the entire state as the community it serves, while a locally funded agency may serve only residents of a given county, city, or even neighborhood. The shift in delivery systems from larger institutions to smaller, community-based organizations has created an increase in the number of agencies that serve fairly local populations as compared to the number of larger institutions that serve broader geographic areas (Mehr, 1998).

Capturing information about your agency's community will be a multifaceted, complex task. Information about the community can be found in a multitude of places, but all of the information probably cannot be found in any one location. One of the best places to look for information about the community is in a recently conducted community needs assessment. Needs assessments of this type are conducted by numerous organizations including the United Way, the local hospital, the local community college, police departments, health departments, and/or the county and city government planning offices.

Extensive information is also available in U.S. Census reports. The website for the United States Bureau of Census provides excellent and detailed information about your community. The U.S. Census Bureau information also enables you to see your city, region, and state in its national context. For example, you can find information that details education levels, income, and general demographic data in your county as compared to those in other counties and states.

State and local government websites are also good sources of information about your community. Additionally, many special interest groups offer websites with relevant information. For example, the Alliance for Families and Children has an extensive website that offers state-by-state data on various measures of child and family well-being, comparing local data with national figures.

Where no written, formal data exist, you might draw upon the impressions of long-standing community members. For example, a person who has worked in your agency for some time will probably have information or at least impressions about which particular geographic areas are heavily served by your agency, even if no hard data exist on the topic. Discussing these questions with one or more professionals in the community or in your agency will begin to expand your understanding of the community.

◆ **EXERCISE 3.9 ANALYSIS**

Following are some key questions to answer as you get to know the community your agency serves.

What relevant demographic information is available about the community? Gather statistical information about such characteristics as income, education levels, unemployment rates, age distribution, and racial and ethnic group distribution within the community.

What racial and ethnic groups are prevalent in the community? What special challenges are presented by cultural diversity in the area?

What is the general character of the community served by your agency: Urban, suburban, rural? How does this affect your agency's service delivery?

What are the prevalent social problems and unmet needs in the community at the present time?

What are the major strengths or assets of the community? What are its major weaknesses or limitations?

Knowing the demographic profile of the area your agency serves is imperative in evaluating how effectively the organization is reaching the community. All agencies must be concerned with how well they are serving various components of the community. Recognizing *overrepresented groups* and *underrepresented groups* is a critical step in this analysis. Social problems do not affect all segments of the community equally, so no organization should expect that its client profile will mirror that of the community on the whole. Nevertheless, if numbers are highly skewed, this may be a red flag indicating obstacles to accessing the service.

In one local children's mental health program, for example, data reflected that one particular school zone was very much underrepresented in the referrals the agency received. The mental health program staff responded to this observation by extending themselves to the school counselors and school social workers in that school zone. Getting to know one another one-to-one, discussing the mental health services available, and hearing about particular children the school staff were concerned about ultimately led to this area of the county being better served by the mental health program.

In a family services center, the staff recognized that Latino families were very underrepresented. Although the agency did serve some Latino families, only about 3 percent of the client population was Latino. In contrast, the community's demographic composition was 21 percent Latino. Clearly, although the Latino population was present, it was being underserved.

In order to serve the community optimally, it is equally important for agencies to recognize *overrepresented groups*. Staff in a children's services organization recognized that many of their referrals came from a particular part of town and that most of those referrals involved substantiated cases of abuse and/or neglect. Approximately 20 percent of referrals came from this one neighborhood although the neighborhood housed only about 5 percent of the town's population. The staff decided to do some prevention work in that community and approached clergy in the area to see if they might provide some parenting workshops in their congregations.

Similarly, juvenile court services noted that African-American males were over-represented in their client population. More than 50 percent of the juvenile cases were African-American males although only 17 percent of the local population was African-American. In response to this observation, court counselors initiated discussions with the local police to learn more about how they monitored and made decisions regarding youth offenders. Through these discussions, efforts were made to reduce race-related bias in the processes used. Also, juvenile court counselors began working with school officials and other community organizations to offer stronger mentoring programs for high-risk youth.

As these examples illustrate, overrepresentation and underrepresentation are concepts that can be applied to clients in particular geographic locations as well as to clients in various demographic groups. In either case, the data are useful in providing high-quality services in the community. Barriers to service access can be addressed and at-risk populations can be targeted for prevention services and other community interventions. The first step of this process is simply identifying overrepresented and underrepresented groups.

◆ EXERCISE 3.10 ANALYSIS

What geographic areas of the community are overrepresented or underrepresented in the agency's client population, if any? What ideas are there about why this might be the case? What methods might be used to more effectively reach and serve the under-served geographic areas?

What racial or ethnic groups, if any, are overrepresented or underrepresented in the agency's client populations? What ideas are there about why this might be the case? What methods might be used to more successfully reach and serve under-served groups? What methods might be used to better serve the overrepresented group(s)?

The information that you have gathered about your community may lead you to discover specific areas in your knowledge and skills that need development. For example, if your community includes ethnic or racial groups that you are not knowledgeable about, it would be helpful to read and learn more about these cultures before working with them. Similarly, being aware that you are working in a community with high unemployment can help prompt you to learn more about resources that address this problem.

Understanding the Agency's Legal, Social, and Political Contexts

Kerson (1994) suggests that the laws that affect practice in a given setting and the prevailing social and political attitudes surrounding the work of a given organization strongly influence that organization's functioning. More formalized aspects of this context are sometimes collectively referred to as *social policy*. Social policy can be defined as "a set of carefully chosen guidelines that steer present and future decisions" (Halley et al., 1998, p. 97) regarding the well-being of people within a given society. Because it guides program and service development, existing social policy affects how human service professionals intervene with clients and address human problems. Less formalized aspects of the social context for human service delivery include the attitudes, values, and opinions that are prevalent within a given community.

Within the U.S. political system, there is a strong relationship between formation of formal social policy and these less formal dimensions of public opinion. Public opinion fuels political interest in certain social problems and dampens interest in others. Similarly, prevailing values and attitudes about certain social problems influence the shaping of social policy surrounding those problems. Thus, social policy and public opinion function as dynamic, interrelated forces that strongly influence the human services delivery system and its various components.

Such issues are interesting and important factors in understanding your field site; they form the larger backdrop for all of the agency's activities. Despite their importance, these issues might be particularly challenging to identify because they may not be obvious in the day-to-day routines of the agency. In examining the legal, social, and political context of your agency's work, you might think of the analogy of looking at your agency through a wide-angle lens rather than through a microscope. In doing so, you will think not so much about the agency's functioning in and of itself, but about the legal, social, and political forces at the local, state, and federal levels that affect practice and policy within the organization. To gain insight into social policy issues affecting your client groups check out the website for the Center on Budget and Policy Priorities. There you will find discussion and analysis of national issues affecting numerous client groups.

Some examples may be helpful in illustrating this level of agency context. On the federal level, major shifts in social policy have occurred in recent years regarding how to respond to the problem of poverty in our society. States are now required by the federal government to develop plans for reducing the number of individuals receiving Temporary Assistance for Needy Families, commonly referred to as "welfare." This policy, of course, affects the activities and climate within public welfare agencies, and it also affects a number of other types of agencies as well. For example, "reducing the welfare rolls" can increase client demands for other services, such as literacy programs, homeless shelters, childcare, and crisis services.

Such increased demands inevitably affect the day-to-day work of the organization as well as staff workloads. Prevailing values and attitudes that suggest that the poor should just "pull themselves up by their own bootstraps" can make it difficult for these agencies to garner the support and funding that they need in order to do their work effectively. In addition, the prevalence of such attitudes in the community and in society at large can create a climate in which the human services professionals who serve the poor feel devalued and unappreciated, creating additional stress and discouragement within these organizations.

On the local level, the attitudes and values of a given community can strongly affect the types of human services initiatives that are and are not supported. For example, a community in which gay males and lesbians are more accepted might provide greater support for efforts to assist HIV patients or gay and lesbian parents than might a less accepting community. The demographic composition of the community clearly plays a role in this phenomenon. For example, in communities largely populated by older adults, the prevailing attitudes might be less supportive of services for children; while in communities dominated by young families, the opposite might be true.

At any given time, particular social issues seem to gain prominence in the national discussion. Certain attitudes can sometimes seem clearly dominant in these discussions, such as the current strong public feelings about immigration. In other situations, however, there seems to be more debate than consensus because the attitudes and values represented in the discussion are more divergent. There have been lengthy and ongoing debates, for example, on such issues as the nature of sex education curriculum in the schools, contraception availability to minors, methods of curbing teen violence, methods of mainstreaming children and adults with disabilities into the community, humane treatment of the elderly, treatment of people with mental illness within managed care plans, and so on.

Reading newspapers and news magazines will help you to stay abreast of important themes and ideas in the ongoing national discussion of various social problems. As a human service professional, these discussions can become very important to your work and to the clients you serve. Therefore, it is essential to pay attention to these issues and contribute to the discussions, particularly as they relate to clients' needs.

Participating in the political and legal processes involved in policymaking are important aspects of professionals' ethical responsibilities according to the Ethical Standards of Human Service Professionals (2005):

> Human service professionals are aware of local, state, and federal laws. They advocate for change in regulations and statutes when such legislation conflicts with ethical guidelines and/or client rights. Where laws are harmful to individuals, groups or communities, human service professionals consider the conflict between the values of obeying the law and the values of serving people and may decide to initiate social action.

A Student's Reflections on the Social Context of the Organization

Working in the area of teen pregnancy prevention is really tough and frustrating in a system that does not allow the discussion of contraception and requires that professionals only discuss abstinence with kids. In addition to this official policy, the school I am in is in a very rural part of the county where most of the students' families hold very conservative views. Even if the policy were different, there would still be a lot of pressure in this community to maintain the same system. As a result of this situation, it seems that the counselors' hands are often tied when it comes to helping the kids consider their options.

◆ EXERCISE 3.11 ANALYSIS

What social policy issues relate to your agency's work? What ideas, attitudes, and values about particular human and social problems influence the manner in which your agency goes about its work? In what ways is your agency's work encouraged and/or discouraged by these attitudes and values?

What specific state and/or federal policies and/or laws affect your agency's work? What prevailing philosophies, values, and attitudes are currently guiding policymaking as it pertains to your setting and the population it serves?

How do the values and attitudes of the local community impact your agency's work? How are these values and attitudes shaped by the community's demographic composition? For example, how might such factors as age, level of education, or political orientation of the residents of the community impact its values and attitudes and thus impact your agency's work?

Conclusion

This chapter has introduced you to a number of ideas and strategies for getting to know your agency, encouraging you to go beyond the more obvious features of the organization's identity. Because the agency has been shaped and continues to be shaped by a host of internal and external factors, these factors have been considered here. The agency's staff, mission, administrative structure, clients, and funding clearly play major parts in its identity and character. Also, the larger human services system in the community influences the agency's identity as it occupies a specific niche within that system. The population of the community and the values and attitudes it holds have a significant impact on human service programs and services. Legal, social, and political issues at the local, state, and national levels set the parameters for the human services system and each organization within it. A thorough understanding of your organization, based on the discussion and exercises in this chapter, will give you greater insight into the various experiences you will have during your internship.

◆ FOR YOUR E-PORTFOLIO

What particular elements of agency context have you learned most about through your internship? How would you describe your agency's context? Consider the organization's community context as well as its national, legal, and political contexts. How has your view of agencies and their contexts changed since the beginning of your internship?

References

Alle-Corliss, L., & Alle-Corliss, R. (1998). *Human service agencies: An orientation to fieldwork*. Pacific Grove, CA: Brooks/Cole.

Barker, R. (1996). *The social work dictionary*. Washington, DC: NASW Press.

Brueggemann, W. (2006). *The practice of macro social work* (3rd ed.). Belmont, CA: Thomson.

Dobelstein, A. (1978). Introduction: Social resources, human need, and the field of social work. In A. Fink (Ed.), *The field of social work* (pp. 3–21). New York: Holt, Rinehart & Winston.

Ehlers, W., Austin, W., & Prothero, J. (1976). *Administration for the human services: An introductory programmed text*. New York: Longman.

National Organization for Human Services. (2005). *Ethical standards of human service professionals*. Retrieved December 15, 2005 from http://www.nohse.org/ethics.html.

Garner, J. (1995). Long-term care. In R. Edwards & J. Hopps (Eds.), *Encyclopedia of social work* (pp. 1625–1634). Washington, DC: NASW Press.

Gibelman, M. (1995). Purchasing social services. In R. Edwards & J. Hopps (Eds.), *Encyclopedia of social work* (pp. 1998–2007). Washington, DC: NASW Press.

Hagen, J. (1995). JOBS program. In R. Edwards & J. Hopps (Eds.), *Encyclopedia of social work* (pp. 1546–1552). Washington, DC: NASW Press.

Halley, A., Kopp, J., & Austin, M. (1998). *Delivering human services: A learning approach to practice*. New York: Longman.

Kerson, T. (1994). Field instruction in social work settings: A framework for teaching. In T. Kerson (Ed.), *Field instruction in social work settings* (pp. 1–31). New York: Haworth Press.

Lauffer, A. (1984). *Understanding your social agency*. Beverly Hills, CA: Sage.

Long, L., & Doyle, M. (2004). Human services: Necessary skills and values. In H. Harris, D. Maloney, & F. Rother (Eds.), *Human services: Contemporary issues and trends* (3rd ed., pp. 67–76). Boston: Allyn & Bacon.

Mehr, J. (1998). *Human services: Concepts and interventions strategies*. Boston, MA: Allyn & Bacon.

Neukrug, E. (2004). *Theory, practice, and trends in human services* (3rd ed.). Pacific Grove, CA: Brooks/Cole.

Ortiz, L. (1995). Sectarian agencies. In R. Edwards & J. Hopps (Eds.), *Encyclopedia of social work* (pp. 2109–2116). Washington, DC: NASW Press.

Russo, J. (1993). *Serving and surviving as a human-service worker.* Prospect Heights, IL: Waveland Press.

Scott, J., & Lynton, R. (1952). *The community factor in modern technology.* Paris: United Nations Educational, Scientific, and Cultural Organization.

Sweitzer, H., & King, M. (2004). *The successful internship* (2nd ed.). Belmont, CA: Brooks/Cole.

Thomlison, B., Rogers, G., Collins, D., & Grinnell, R. (1996). *The social work practicum: An access guide.* Itasca, IL: Peacock.

Weiner, M. (1990). *Human services management: Analysis and applications.* Belmont, CA: Wadsworth.

Chapter 4

Learning to Learn from Experience: The Integrative Processing Model*

After spending a week or two in their field agencies, most students begin to experience increasing responsibilities and a greater sense of independence. They often feel that they are starting to "test their wings" as human services workers. Also, many students begin to enjoy their fieldwork more, finding that they are busier and able to make more of a contribution to their organizations.

Despite these positive developments, a word of caution may be necessary. Being busy does not ensure that learning is going on, nor does enjoying your work site and co-workers, nor does finding the work interesting. Learning from experience is a complex process that takes intentional effort, engaging and combining your intellect, behavior, and emotions. This chapter helps you understand more about the processes involved in learning by doing and introduces to you a model for learning from experience that can enhance the quality of your thinking, learning, and performance in the field.

The Role of Reflection in Learning from Experience

Experience, in and of itself, does not necessarily teach very effectively. Your experiences in your fieldwork must be thought about, reflected upon, and analyzed to yield the greatest learning (Bogo & Vayda, 1995; Hutchings & Wurtzdorff, 1988; Kolb,

---

* Adapted and used with permission of *Human Service Education*. From P. Kiser. (1998). The integrative processing model: A framework for learning in the field experience. *Human Service Education 18*(1), 3–13.

1984). You might think of experience without reflection as being somewhat like an unread book sitting on a shelf, or perhaps a book that you have quickly scanned but not carefully read, understood, or thought about.

A primary objective of your human services fieldwork is developing the ability to make meaningful connections between your practice experience and your knowledge so that your work is guided and informed by your knowledge of the field. If you are like many students, you might find this surprisingly difficult to do. A number of factors contribute to this difficulty. You might have studied relevant content and skills some time ago, and as a result, you feel rusty. As days become busier, sometimes even hectic, you may feel that "there isn't time to think" and only barely enough time to do what must be done. Another significant factor is that after years of theoretical, intellectual work in the classroom, it feels good to be busy, active, and on the go.

Falling into an unreflective way of working, however, is a trap that will severely limit what you learn as a student and your effectiveness as a professional. If you find that you are truly too busy to engage in reflection and planning, discuss this with your field supervisor. Your responsibilities may need to be changed so that you can do fewer things with higher quality.

Most of the time, students do have sufficient time for the integration of knowledge and theory with practice, but they need to set aside time for in-depth reflection and possibly even for reading and reviewing relevant content. If you have saved your textbooks and notes from previous coursework, these will serve you well for such reflection and review during your internship. Additionally, students often need some guidance in *how* to do this integration of theory and practice. In essence, students generally need to learn how to learn from experience most effectively.

STUDENT EXAMPLE

George was a student intern in a day treatment program for troubled adolescents. Within the first week or two of his placement, staff members observed that George was overly controlling and critical toward the teens in the program. After about a week of observing such interactions and encouraging him to try other methods of interaction, staff members became frustrated with George's lack of progress and asked him simply to observe for the next week, not interacting with the clients at all.

Ironically, George had a 4.0 grade point average and could no doubt earn an "A" on any written test regarding theoretical information, including principles governing the development of positive helping relationships. His knowledge of theory, however, was of no use because he seemed to be unable to convert this knowledge into effective practice.

During his week of "observation only," George's supervisor asked him to reflect on what he had learned in the classroom about helping relationships and to explore his own beliefs, assumptions, and feelings about the teens in the program. Although George was angry and defensive about the feedback he had received from the staff, he tried to respond to the questions his supervisor had raised by writing honestly in his journal each day, focusing on the issues assigned.

After a few days, George and his supervisor reviewed what he had written and discussed it. In this discussion, George's supervisor was able to help him identify some erroneous assumptions, misinformation, and stereotypes that were contributing to his angry and controlling feelings toward the clients. George's supervisor was able to give him more accurate information about the teens in the program as well as about the staff. Although he continued to struggle some with his feelings, George was better able to use his academic knowledge in the subsequent weeks of his internship.

◆ EXERCISE 4.1 PERSONAL REFLECTION: OBSERVATION OF SELF AND OTHERS

What could George have done earlier in his internship to enable him to use his knowledge more effectively?

Reflect on your own internship. Try to identify any occasions in your own experience in which you might have used your knowledge more effectively. Also try to identify the obstacles that interfered with your ability to apply what you knew in these situations.

A Student's Reflections on Learning from Experience

I am so busy now that I seem to roll straight from one event to another all day long. My supervisor has also been busy, so it is hard to even find the time to talk at any length. He is usually pretty reliable in making time for me though. I always think

continued

---

that I am learning as much as I possibly could, but then when we sit down together, he is always able to point out so much more that I didn't see before. It is the same with writing in my journal. The other night when I wrote in my journal I described a discussion I had observed between two of the workers on the unit. In the process of writing, I realized the two different theoretical perspectives (humanistic versus behavioral) that each of the workers was coming from in his point of view. I hadn't made the connection between the points they were making and the theories I had studied until then. It was neat to have this insight because it gave me more understanding of what they were talking about. It also made me proud of myself that I had put it all together without someone having to point it out to me.

An Overview of the Integrative Processing Model

Much has been written about how students can best learn from experience. In this chapter, I propose a six-step, cyclic process, which draws in part upon the strengths of various models that have been created about learning from experience. I refer to this model as the Integrative Processing Model (see Figure 4.1).

The model is described as *integrative* because it calls upon you, as a fieldwork student, to draw upon many components of yourself and your knowledge to extract maximum learning and meaning from your experiences. During your fieldwork, the Integrative Processing Model will serve as a tool to help you reflect upon and think through your field experiences carefully and systematically, integrating major components from your education, including knowledge of theory and content, self-awareness, and professional ethics. Your knowledge, behavior, attitudes, emotions, and values all come into play as you process your experiences using this model. The Integrative Processing Model consists of the following steps: (1) gathering objective data from the concrete experience, (2) reflecting, (3) identifying relevant theory and knowledge, (4) examining dissonance, (5) articulating learning, and (6) developing a plan.

This model is equally useful to those students working in direct services to clients and to those working in administration and other indirect services. You may use it as a structure for written reflection and analysis of your experiences or as a purely cognitive process. The model provides an excellent structure for writing meaningful,

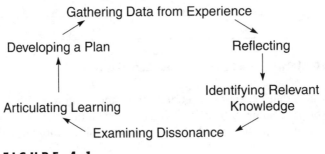

FIGURE 4.1
The Integrative Processing Model

constructive journal entries, and it is recommended that you use it as such. Write your responses to each step in the six-step model, processing several experiences each week in your journal. Writing your responses will help you to think through each step of the process thoroughly and carefully.

Because of the autonomy that students typically experience in their fieldwork, the Integrative Processing Model is designed for use as a somewhat self-guided process. It is assumed, however, that you will use it under the supervision of your field supervisor and/or your faculty supervisor. The field seminar provides an ideal opportunity to learn the model and practice using it as you analyze and discuss your field experiences with other students. The model also provides an excellent structure for discussing specific experiences with your supervisor during supervisory sessions. Over time, you will be able to use the model with greater autonomy and ease, applying it to everyday professional practice. With repeated practice, you will internalize the model as a matter of habit, using it as an ongoing method for thinking about and learning from practice experiences. Thus, using the model routinely can enhance your learning, not only as a student during your field experience but also subsequently as a practicing professional.

Each step of the Integrative Processing Model is discussed in some detail in this chapter. Following the description of each step, you are asked to apply that particular step of the model to a situation that you have encountered in your fieldwork. Before reading the following sections, select an experience that you have had during your field placement that you can use as your practice example. It is most useful to select an experience that you found particularly interesting, challenging, troubling, or educational. If you find it difficult to implement a particular step in the model, it might be helpful to read the case examples later in the chapter. These examples give you clear illustrations of how to apply each step.

A Student's Reflections on Processing Internship Experiences

One of the things that I have experienced during my internship is a shift in the way I think and work. I can summarize it best by saying that I have learned to put thinking and working together. In the beginning I was truly flying by the seat of my pants. Now I am much more focused, deliberate, and intentional in what I'm doing. Although there are always some curve balls that come along, I can usually explain why I handled a situation the way I did. I think about what I know (at least sometimes) and can actually use it to give me some ideas about what to do next or how to approach a situation. It is a good feeling to have all of these things start to come together.

Step 1: Gathering Objective Data from Concrete Experience

Models of learning from experience often suggest that learning begins with experience itself (Bogo & Vayda, 1995; Kolb, 1984). In your fieldwork, experience forms the basis for learning because each specific experience is an event from which learning must be extracted. The experience may be one in which you are an active participant or an observer. In subsequent steps of the Integrative Processing Model, this experience becomes the focus for reflection and application of knowledge. During the experience, as well as after the fact, you are gleaning information about the situation and about the behaviors, actions, and/or interventions of the various participants. Your ability to be an objective observer of experience is developed through this step of the learning process. In addition to your own observations, you might also use additional methods of observation.

For example, you might arrange for your supervisor to observe an upcoming experience and provide an objective view of the situation. You might (with the permission of any other participants) audiotape or videotape the interaction. Writing process recordings in which you re-create the dialogue (both verbal and nonverbal) between yourself and others as precisely as possible is another good method of gathering objective data. An additional advantage of this approach is that it helps you to develop your own skills as a participant-observer—that is, to engage actively in the interaction as well as to stand outside the interaction psychologically and watch it. Objective information, retrieved from the concrete experience by a variety of methods during Step 1, provides the focus for subsequent steps in the Integrative Processing Model.

◆ **EXERCISE 4.2 PERSONAL REFLECTION: OBSERVATION OF SELF AND OTHERS**

Select an experience to use throughout this chapter to practice the various steps of the Integrative Processing Model. In this first exercise, consider the methods of identifying objective information discussed above. Using these methods, try to report objectively the important events, information, and facts from the experience you have selected and record those observations here.

Step 2: Reflecting

One of the most widely agreed upon precepts of learning from experience is the importance of reflection, as was discussed earlier in this chapter. In Step 2 of the Integrative Processing Model, you engage in personal reflection by assessing your own reaction to the situation. You might think of this step as examining your "involvement of self" in the task at hand. How does this situation touch upon your own values? How does it relate to your personal history or similar experiences that you might have had? What thoughts and emotions does this situation trigger in you? What assumptions are you making about the situation? What assumptions are you making about the people involved in the experience, including yourself?

Another important component of reflection is evaluation of your own behavior in the concrete experience. As you examine your verbal and nonverbal behavior in the situation, what behaviors enhanced your effectiveness? What behaviors diminished your effectiveness?

Human services education emphasizes that professionals must develop and maintain a high degree of self-awareness in order to function effectively in the field. The reflection required in Step 2 helps to raise your awareness of the feelings, attitudes, behaviors, values, and assumptions that you are bringing to the particular experience.

 ◆ **EXERCISE 4.3 PERSONAL REFLECTION: OBSERVATION OF SELF AND OTHERS**

Using the preceding discussion as your guide, discuss your personal reaction to the situation. Be sure to consider the various components of reflection discussed previously in formulating your response.

Step 3: Identifying Relevant Theory and Knowledge

Students engaged in fieldwork that is linked with professional preparation must make connections between what they have learned in the classroom and what they experience in the field (Argyris & Schon, 1974; Bogo & Vayda, 1995). Step 3 of the Integrative Processing Model requires you to identify theoretical, conceptual, and/or factual information that can shed light on requires your to fieldwork experience. The information you have recorded in the previous steps of the model may constitute only a set of relatively meaningless, disjointed facts if examined outside the context of relevant theory and knowledge. Although previous classroom learning forms the foundation for Step 3, you might also need to engage in more extensive reading and research to expand your knowledge.

For example, if you are working in an administrative role during your field experience, you might draw upon your existing knowledge of organizational theory and leadership theory but find that you need to do additional reading about models of supervision and strategic planning. If your fieldwork involves direct services to clients, you might draw upon your knowledge of theories regarding human behavior, the dynamics of the helping relationship, and the stages of the helping process, but feel the need to do additional reading about a particular client's presenting problem or a particular issue in family dynamics.

Against the backdrop of relevant knowledge, certain information identified in the previous steps may rise to the foreground in its importance whereas other information might become relatively less significant. Some facts may begin to cluster together, bearing some relationship to one another, forming a more cohesive picture, pattern, or theme. The application of knowledge—theoretical, conceptual, or factual—provides an organizing focus, a lens through which you can view and make sense of the experience.

◆ EXERCISE 4.4 SYNTHESIS: LINKING KNOWLEDGE AND EXPERIENCE

Using this discussion about applying relevant knowledge as your guide, identify specific academic information that might be relevant to the situation you have selected. Consider theories and concepts that you have studied as well as any research or factual material that you are familiar with that relates to the situation. Discuss how this information contributes to your understanding of the experience.

Step 4: Examining Dissonance

Some authors advance the idea that dealing with dissonance is an important component of learning from experience (Argyris & Schon, 1974; Hutchings & Wurtzdorff, 1988). Dissonance is defined as "the lack of harmony, consistency, agreement" (*American Heritage Dictionary*, 1990, p. 409). Step 4 of the Integrative Processing Model suggests that, having examined the experience itself, your personal reactions, and relevant knowledge (Steps 1–3), you are now in a position to explore points of dissonance in the situation.

Dissonance can exist on a number of levels as you react to your experiences. You might experience intellectual dissonance as competing theories offer divergent points of view (Step 3) or as conflicting data arise out of the concrete experience (Step 1). You might experience dissonance between the espoused theories of the profession and your own personal views. You might find that your thoughts and feelings clash (e.g., "I know I am supposed to be warm and empathic toward my clients, but I really don't like this person."). Conflicts might also occur between theory and practice ("I did what we discussed in class, but it didn't work") and between your thoughts and behavior ("I know what I am supposed to do, but I just can't do it right" or "I know what I am supposed to do, but I just can't make myself do it"). Following are some guiding questions that can help you to identify points of dissonance in your work:

- What, if anything, do I feel uncomfortable about in this situation?
- What disagreement is there between what I "should" do and what I "want" to do?
- What mismatch is there between what I "should" do and what I "must" do?
- What conflict is there between competing "shoulds" in the situation?
- What disagreement is there between my personal views of the situation and views offered by the theories and knowledge of the profession?
- What conflict is there between what I "know" and what I "do"? (Hutchings & Wurtzdorff, 1988)

Becoming clearly aware of dissonance is the first step in its resolution. Once recognized, such dissonance can often be resolved at the time it emerges. The values and ethical principles of the profession in many instances provide a useful framework for resolving dissonance. Values such as client self-determination, respect for the client's culture, recognition of client dignity and worth, appreciation of individual uniqueness, confidentiality, and nonmanipulative intervention are examples of useful guiding principles in reconciling conflicting points of view.

Some points of dissonance, however, are more difficult to resolve. Particularly challenging are those situations that involve difficult ethical dilemmas. Such situations need careful consideration and consultation. (Chapter 8, Developing Ethical Competence, deals with this topic in greater detail.) Other issues might present difficulty because they intersect with your strongest personal values or demand change in deeply ingrained assumptions or habits. These more troublesome issues may need to be discussed and considered for some time before resolution is achieved. If they present obstacles to your working effectively, these points of dissonance should become a central focus in your supervision.

It is also realistic to point out that at times the dissonance that you experience may not be reconcilable. You might find that rather than reconciling dissonance, you must instead learn to operate within it. In such instances, professionals acquire the ability to tolerate ambiguity, to embrace competing values, to come to terms with their own personal limitations, and even to come to terms with the limitations of their profession.

Dissonance may not be involved in every experience, but it does occur frequently. Although the issues confronted in this step of the model are difficult and often uncomfortable, dissonance should be viewed as a necessary and desirable part of learning within the field experience. Through confronting and struggling with difficult issues, you will experience some of your most significant and meaningful learning. During this step, your total self is brought to bear upon the learning experience as your knowledge, skills, personal reactions, and values are wrestled with and brought to some degree of congruence with one another. Through this process genuine learning occurs, and professional maturity develops.

♦ **EXERCISE 4.5 ANALYSIS**

Using this discussion about dissonance as your guide, identify points of dissonance that you experience(d) in the situation you have described in your practice example. Discuss the specific aspects of this dissonance that you need to work on resolving in order to gain greater clarity, comfort, and confidence in your work. Also identify any particular points of dissonance that might seem to be irreconcilable.

Step 5: Articulating Learning

Students often report with excitement that they are "learning so much" from their fieldwork, but when specifically asked what they have learned, they all too often fall silent or lapse into vague generalizations. This scenario perhaps reflects the fact that, although students might indeed have learned a great deal, they have not thought carefully enough about this learning to be able to put it into words. Step 5, Articulating Learning, requires you to put your learning into words. Using words to explain and describe what you have learned pushes you to conceptualize your learning. Through writing, a dim awareness can become clarified into a coherent statement. Once you have constructed this statement, the knowledge becomes more clearly your own. You then have greater command over this learning as a more tangible, concrete, and lasting "possession" that you can retrieve and use as needed.

The guiding question in this step of the model is straightforward: "What are the major lessons that I can take from this experience?" The lessons learned may have to do with skills that you developed, theoretical knowledge that you gained, insights that you developed into yourself or others, or deeper understanding that you acquired of an ethical principle. Whatever the lessons, remember that all knowledge is tentative rather than fixed and immutable. You must test the learning gained from a given experience in subsequent experiences. In doing so, the lesson may be reinforced, refined, revised, or refuted. Through this process, you not only grow in your knowledge and skills but also acquire the skills and habits of an active learner, equipped to continue learning from experience throughout your career.

A Student's Reflections on Articulating Learning

A new volunteer came in today to start her service hours for a course she is taking at the college. I was asked to spend some time with her and explain how things work around the agency. When I tried to explain certain things to her, I sometimes found that I wasn't really as clear about it in my own mind as I thought I was. I had to go ask my supervisor and other staff members questions just to get things straight. As a result, I learned some things from the process of orienting the volunteer. Until I had to put it into words, I didn't even realize what I didn't know.

 ◆ **EXERCISE 4.6 ANALYSIS**

Using the discussion about articulating learning as your guide, identify the major lessons that you are taking away from this experience.

Step 6: Developing a Plan

The final step, Developing a Plan, is a two-pronged step of the process in that it calls upon you to think through (1) how to proceed in your work (Bogo & Vayda, 1995) and (2) how to proceed in your own learning and development. Each of these topics is considered in this section.

The work you have done in Steps 1–5 provides a solid foundation for making an informed choice as to how to proceed in your work. In some situations, the work of

Steps 1–5 may lead you to a clear action path, but in many situations, a number of alternatives will be available. Step 6, Developing a Plan, consists largely of decision making through identifying, evaluating, and selecting from the various alternatives.

Ideas for alternative courses of action emerge in large part from Steps 3 and 5 of the Integrative Processing Model. Application of knowledge in Step 3, Identifying Relevant Knowledge, yields implications for practice, generating ideas for potential courses of action. Learning identified in Step 5, Articulating Learning, may also hold implications for what might be done in the next experience. Although Steps 3 and 5 are particularly helpful at this point, the work done in any or all of the steps of the Integrative Processing Model may offer ideas for additional options as well as support for or reservations about certain options. Additionally, common agency practices often suggest or reinforce certain options, while agency policy and protocol more clearly direct or even dictate the next step in some situations.

In Step 6, identify and weigh each alternative plan with consideration of the following: (1) What are the likely consequences of this plan? (2) What factors and forces support the selection of this plan and the likelihood of its effective implementation? (3) What factors and forces argue against the selection of this plan or present obstacles to its effective implementation? Out of these considerations, you (in conjunction with your field supervisor) will select a specific plan for your next experience with the task or assignment.

For the development of your own autonomy, it is best for you to function as independently as possible in developing your plan. Submitting written summaries of your experiences in which you work through each step of the Integrative Processing Model to your supervisor is beneficial in this regard because it ensures that you think independently, get feedback before you proceed with your plan, prepare adequately for your supervisory conferences, and use your supervision time in a focused, efficient manner. Due to time constraints alone, it is impossible to discuss thoroughly every experience that you have in your fieldwork with your supervisor. Written summaries enable you and your supervisor to identify the most difficult situations for close review in your supervisory sessions, whereas less difficult situations might be more quickly reviewed.

In addition to developing a plan for your future work, an equally important element in this step of the model is developing plans for your own learning. As you work through the steps of the model and make plans for subsequent work, you will more clearly recognize gaps in your knowledge and skills that need to be addressed. Developing a plan for your future work must include an assessment of whether you currently have the knowledge and skills required to implement that plan successfully or whether you need to acquire that knowledge and those skills. A plan for your learning might include any number of activities: reading, observing another worker, attending a workshop or seminar, or doing a role play with your supervisor.

Once plans are developed for your next experience with the task or project, you are poised to begin the learning cycle again. Even in cases in which you will have no further immediate opportunities to work with the particular client, project, or task, you should still complete the final step by developing a plan. In doing so, you should look toward your next concrete experience as the next professional experience in which you will deal with a similar situation. Considering the question, "How might I handle things the next time I have such a responsibility?" will prove to be good preparation for this next experience whenever it should occur. Whether the next concrete experience is the next day or sometime in the future, thinking through your plan forms a basis for your activity in another professional experience. This next experience offers the opportunity to start the learning process again at Step 1, Gathering Objective Data from Concrete Experience.

 ◆ **EXERCISE 4.7 SYNTHESIS: LINKING KNOWLEDGE AND EXPERIENCE**

Continuing to focus on the situation that you have discussed in Steps 1–5, develop a plan as to how you will proceed in your work. Think about where you need to go from here in the immediate situation. If you do not anticipate further work in this immediate situation, think about how you might handle similar experiences in the future.

Finally, and very importantly, identify specific knowledge and skills that you need to gain in order to conduct your future work effectively. Develop a plan for gaining such knowledge and skills.

Applying the Integrative Processing Model

The following examples of students using the Integrative Processing Model should help you understand more clearly how to use the model yourself. The examples also give you some ideas about how to use the model as a structure for written reports.

CASE EXAMPLE 1: A DIRECT SERVICE SITUATION

The following example is adapted from a student's encounter with an elderly client during her internship. The example serves to illustrate how the model helped her to think through the situation carefully and thoroughly, improving both the quality of her learning as well as the quality of the services that she was able to provide the client.

Step 1: Gathering Objective Data from Concrete Experience

I visited Mrs. S., an 89-year-old widow, in her home. Mrs. S. was using a kerosene heater to heat her small, cinderblock home, and it was very hot inside. Her vision is quite poor, and she walks with a metal-frame walker. She does her cooking on a two-burner hot plate. Mrs. S. reported a lengthy story about her children stealing her land from her and wanting to put her in a nursing home. She repeatedly said that she needs a lawyer but can't afford one. She also said, however, that her children don't want her to go to a nursing home because it would take all her money and property. Mrs. S. was tearful and worried about her future. She worries about how she will take care of herself. She reported that she often goes for as much as a week or two without getting out of the house or having a visitor. I listened to Mrs. S. and empathized with her position. I told her I would find out what services might be available to help her stay in her home and that I would see if any legal help could be provided for her.

Step 2: Reflecting

I felt very sad for Mrs. S. She lives in a situation that I would not want one of my family members living in. I felt angry with her children for allowing her to live this way and not offering her more help. When I was talking with her, I bought into the story about her children taking her land because she was crying and so convincing. Thinking about it later, I realize that this might not be true. I do believe there is legitimate reason to be concerned for her safety due to her heating and cooking methods, combined with her physical problems. I have an impulse to try to rush in and fix everything here.

My grandfather lived with my family for the last 12 years of his life because he couldn't function very well on his own. I simply cannot imagine that none of Mrs. S.'s four children can help her out by having her move in with them or vice versa. This makes me really angry and makes me feel very sorry for Mrs. S.

Step 3: Identifying Relevant Theory and Knowledge

From my studies, I know that social support and stimulation are needed to help the elderly maintain optimal mental orientation as well as physical well-being. I have also read numerous articles about elder abuse and exploitation and am aware that they are not uncommon. In fact, children are frequent perpetrators. According to laws in this state, the Department of Social Services must be notified if there is evidence of exploitation so that they can conduct an investigation of abuse and/or neglect.

Although Mrs. S.'s story about her children was convincing, I have also studied dementia among the elderly and am aware that it is not uncommon, especially among the old-old (those over 85). Her beliefs that her children were plotting against her and trying to harm her would fit with the paranoia that is sometimes a symptom of dementia.

Although Mrs. S.'s current living situation is far from ideal in my opinion, I must also remember all of the literature about the harmful effects of institutionalization. For many older people, going into a nursing home or some other kind of group care can contribute to disorientation and physical decline because the individual has less control over his/her environment and possibly less opportunity or need to engage in self-care. All of this should be considered as plans are made for Mrs. S's future.

Step 4: Examining Dissonance

There are many points of dissonance in Mrs. S.'s story. She says, for example, that her children want her in a nursing home and then says they don't want her in a nursing home. It is not clear whether the story about her property being stolen is true. I am uncomfortable with the fact that I told her that I would see if legal assistance might be available. Upon further thought, this might be premature.

There are also a number of points of dissonance that arise in part due to my own reactions to the situation. For example, I feel that it would be best for Mrs. S. to live with one of her children, but she never indicated an interest in this. Also, I am assuming that Mrs. S.'s children are unwilling to help her, but I have not met them and concluded this based only on her report. Although I see Mrs. S. as being in an unsafe living situation, she reported no particular problems living alone except for social isolation. No "close calls" were reported. I have to wonder if my own middle-class lifestyle and values are holding too much sway in my perceptions or is her situation really unsafe. Because I got so wrapped up in my own reactions, it isn't clear to me where Mrs. S. wants to be. I don't think I even asked her directly what she wanted for herself. Does she want to stay where she is? Go to a nursing home? Live with someone? Some other option?

Step 5: Articulating Learning

As I look back at this experience, it is amazing to me how my emotions took the upper hand and how little I accomplished in this contact as a result. I think tears have a powerful effect on me, particularly when they come from an older person. I think my closeness to my own grandparents and my impulses to protect them come into play here. It is hard for me to see Mrs. S. as an adult. I will need to work on this as I go into my next contact with her as well as with other older adults.

Step 6: Developing a Plan

Mrs. S. may be, but is not necessarily, a victim of exploitation. This needs to be explored further. I need to secure permission from Mrs. S. to meet with her children, perhaps in a family meeting. I am not sure I have the skills yet to carry this meeting off. It could be volatile and in any case will require some real skill and tact.

Take a detailed history from Mrs. S., looking closely for any evidence that she has been in high-risk situations due to living independently. Also get the perspective of her children and neighbors on this, if possible. I think I have the skills to do this and should have done so in the first meeting, but I got so wrapped up in her emotions—and mine!—that I didn't think of it. The potential obstacles might be the inaccessibility of her children and neighbors or her possible unwillingness for me to speak with them.

Mrs. S. definitely needs more social stimulation, and she acknowledges her loneliness. Talk with Mrs. S. about possible friends, family, and other supports in her community that we might reactivate. Also see if she is interested in attending a senior center or congregate meal site. This should be easy. I can take some brochures from the senior center. She might be able to see the pictures, and I can read them to her. If this is something she is interested in, she could probably get started very soon.

Try to help Mrs. S. identify what she wants to do, where she wants to live. I can help Mrs. S. explore her options and again probably should have done this in the first contact. I believe it will be challenging due to her emotionality and her tendency to focus on the past. I need to support her feelings but continue to focus and structure the interview on her options at the present time.

In terms of my own learning and development, I think I should role play my next conversation with Mrs. S. There are several things that I need to accomplish in this meeting, and I know how easily I was sidetracked last time. I will ask my supervisor to role play this situation with me and give me any suggestions for conducting this interview more successfully. Also, for my own information as well as possibly for Mrs. S.'s, I need to research the local legal resources for indigent older adults. Whether or not this is called for in Mrs. S.'s situation, my conversation with her has made me aware that I really don't know how such situations are handled.

CASE EXAMPLE 2: AN ADMINISTRATIVE SERVICES SITUATION

The following example illustrates how the Integrative Processing Model can be applied to administrative or indirect service experiences that students encounter in the field. As with the first example, the student's use of the model enhanced his learning as well as the quality of his work.

Step 1: Gathering Objective Data from Concrete Experience

As an intern in the hospital's human resource department, I was asked to organize a series of brief educational seminars on employee wellness. In a meeting with the human resources director, I learned a few facts about the need for the project. He informed me that the department had conducted employee surveys that revealed that two major concerns of employees were "too much stress" and "too little free time." The director had intended to implement some programs to deal with these concerns but did not have sufficient staff available to do so. Having an intern provided an opportunity to develop this long overdue program as well as an opportunity for me to develop my skills in organization and presentation. We decided that I would "kick off" the series with a session on stress management and that the lunch hour would be a good time for the session. Because I had done a presentation in one of my human services classes on this topic, I was well prepared, knowledgeable, and pretty confident.

I planned to spend most of the session presenting information to the group. Because there was not much time available, I needed to use it efficiently to cover the material. I started the session by explaining that the surveys they had completed indicated stress was a concern for them, and I asked them to identify some of their stressors. Almost immediately a number of participants went into lengthy tirades about the administration's personnel practices. A number of long-term employees had recently been "let go" in a drastic cost-cutting measure. Employees complained that those remaining were extremely overworked and fearful for their own jobs. One participant said, "I can't believe this stress-management thing is their response to that survey they did! No offense, but that's just throwing us a bone. We need some real change around here."

We got through about half of the program that I had planned. I did a lot of listening to and reflecting feelings, followed by efforts to redirect attention back to the topic. At the end, I told the group that I hoped they had picked up some useful information but that it mostly seemed that they needed a chance to ventilate. As they left, some participants thanked me and picked up copies of handouts that I had prepared. A couple of participants talked for a few more minutes about the difficult situation at the hospital.

Step 2: Reflecting

I felt really out of control throughout the so-called presentation. From beginning to end, I felt uncertain as to what I should do and fearful that things were going to get even more out of control than they were. I vacillated between feeling angry at the participants for ruining my nicely planned presentation and feeling angry for them because of what they were having to put up with in their working conditions. I kept wanting to say, "Hey, listen, I'm only an intern," but I knew this wouldn't be very professional.

Thank goodness I did remember how important it is to acknowledge and respect people's feelings. This one skill seemed to get me through the hour, although at the time it didn't seem like very much. I am pleased that I was able to maintain my cool and provide at least some support, but I was exhausted when the meeting was over. Also, I felt like I should do something for these people. I felt like marching into the administrator's office and setting him straight. I also found myself feeling angry at the human resources director for assigning me this project. Nobody prepared me for anything more than just a straightforward one-hour educational session. I keep feeling like I must have done something wrong for this to go so poorly. At the same time, however, I recognize that I did the best I could under the circumstances.

Step 3: Identifying Relevant Theory and Knowledge

According to problem-solving theory, once a problem is identified, a number of steps should follow, including gathering information about the problem, generating a number of alternative solutions, and assessing and comparing alternatives. Selecting and implementing a solution should follow these steps. In this case, we jumped straight to a quick fix without going through all the steps to assess what was really needed.

I also have studied in my group dynamics and administration classes that people who are highly skilled in their jobs and highly motivated to perform well tend to function best with leaders who use a participatory style. Authoritarian, or "telling," styles of leadership are most compatible with individuals who have low skills and low motivation. The people whom I was presenting to were people who were professional and highly motivated. Using such a lecture style approach did not fit well with their needs. A program that was based more on dialogue would probably make more sense for them.

Also, the people who work at the hospital are many times in jobs that are at high risk for frustration and emotional overload. People who work in areas in which they experience high psychosocial demands and low control over their environment are at the highest risk of burnout, and this sounded like the kinds of situations they were describing to me in the session. Frustration is one of the stages of developing burnout. Emotional outbursts are typical of this stage in the process. I certainly was hearing some frustration and witnessing some emotional outbursts.

Finally, the information I have learned about triangles in my family class and in my groups and communities class seems very relevant here. Triangling theory suggests that when two parties are in conflict with each other and a third party enters the situation, there is a risk of triangulation. Triangulation occurs when the third party gets stuck in the middle, trying to solve the problem or "take up the banner" for one side of the conflict to challenge the opposite side. I believe I became triangled in a conflict between the administration and the employees when I presented this program.

Step 4: Examining Dissonance

This situation was so uncomfortable that the points of dissonance may be hard to enumerate, but here goes with the biggest issues. I am uncomfortable with the role I am in now. I feel that I need to try to do something helpful, but I am not sure what to do. If I share the employees' concerns with the human resources director or other administrators, I am not sure what the consequences would be. Would the employees get in hot water? Would I get in hot water? I need to discuss this session with the human resources director but I am not at all sure what to say. I am also uncomfortable with the fact that I have agreed to lead more than one of these sessions during my internship. I am not sure this is a good idea now.

I have always learned that the key to making effective presentations is being well prepared, knowledgeable about the topic, and well organized. It seems to me now that this isn't enough.

Step 5: Articulating Learning

I definitely learned more about what it means to be prepared for a presentation. Being prepared means a lot more than knowing your information, having an outline, and being well organized. I need to know as much as possible about my audience and their needs. Although I had thought I was well prepared, important planning had been completely omitted from the process. The most relevant information that I have studied is, I believe, the problem-solving process. I have studied it repeatedly, but I was never aware of the many situations when applying it might be relevant. Now I think I finally get the picture. It will occur to me to use this approach when trying to address more varied situations, and not just accept pat solutions that are offered to me.

Rather than just accept work assignments at face value, I need to ask more questions and exercise more independent thought. Prior to this experience, I also think I had made the mistaken assumption that administrative work was more straightforward and less emotionally sensitive than direct service. Wrong!

I also learned that I can function pretty well on my feet. As I look back at the experience, I believe I handled it pretty well. I could have made the situation much worse by disregarding the feelings that were being expressed by the participants or by letting my own anger and frustration take control. I stayed tuned in to their feelings but also tried to stay task oriented.

Although I have experienced triangulation many times with friends and family members, I now see how easily it can come into play in the workplace, especially since I am the helping type who likes to fight for the underdog. I know that I need to stay sensitive to this issue and make conscious decisions about when and how it is appropriate for me to get involved rather than be led purely by my instincts.

Step 6: Developing a Plan

Before leading another session (if I lead another session), I need to be sure that I clearly understand the nature of the employees' needs and concerns. I am not at all sure the problem-solving steps have really been worked through in this case. I will suggest to the human resources director that we need to gather more information about any identified problems before we try to put together a program to address them. Educational sessions may not be the best alternative. We need to rethink that assumption.

I need to try not to get triangled into the conflict between employees and management. I believe it would be best to acknowledge to the HR director that, as the survey pointed out, the employees were highly stressed and expressed a great deal of frustration in the meeting. I will explain that they seemed to feel more of a need to talk about the stress than to focus on stress-management methods at that particular time. I will suggest, as mentioned earlier, that it would be a good idea for the administration to gather more information about the nature of the employees' stress and concerns so that more precise interventions can be designed to address their needs.

If I do lead additional sessions in the future, I will use a more participatory style of leadership rather than a more formal presentation, or telling, style. This style is more appropriate for leading groups of highly skilled, highly motivated people and will allow for the expression of feelings that I now know are pretty intense. As a result, I will need to prepare less information and be prepared to be flexible.

Also, as I interact with the employees, I will have a greater understanding of their unique work situation, recognizing that it is one that frequently leads to burnout among helping professionals. This will help me not to personalize any reactions I might get, such as short tempers, frustration, or even apathy.

Conclusion

Continue to use the Integrative Processing Model daily in your fieldwork. On most days, you will have at least a few, and possibly many, experiences that you can process using this model. Using the model to help you think through your experiences will create deeper and richer learning for you by ensuring that you engage in thorough reflection and critical thinking.

The Integrative Processing Model provides an ideal structure for discussing significant events in your journal. As you develop more skill in using the model, you can become less structured in your approach, working through it mentally rather than on paper. Until then, continue to write your thoughts on each step. Even after you become more proficient in using the model, you will continue to find it helpful at times to write down your thoughts as you sort out more complex and challenging situations. With practice, you will find that this process begins to feel natural, comfortable, and helpful.

 ◆ FOR YOUR E-PORTFOLIO

As you apply the Integrative Processing Model (IPM) to your work, select at least one IPM write-up to submit to your e-portfolio. Select a write-up that you feel reflects your best work. A strong IPM paper will demonstrate your strengths in observation, self-awareness, application of knowledge/theory, critical thinking, and decision making. In the space below, you might identify a few experiences you have had in your internship thus far that would be particularly interesting to think through using the Integrative Processing Model.

References

American Heritage Dictionary. (1990). Boston: Houghton Mifflin.

Argyris, C., & Schon, D. (1974). *Theory in practice: Increasing professional effectiveness.* San Francisco, CA: Jossey-Bass.

Bogo, M., & Vayda, E. (1995). *The practice of field instruction in social work: Theory and process.* Toronto: University of Toronto Press.

Hutchings, P., & Wurtzdorff, A. (1988). Experiential learning across the curriculum: Assumptions and principles. In P. Hutchings & A. Wurtzdorff (Eds.), *Knowing and doing: Learning through experience* (pp. 5–19). San Francisco: Jossey-Bass.

Kolb, D. (1984). *Experiential learning: Experience as the source of learning and development.* Englewood Cliffs, NJ: Prentice-Hall.

Chapter 5

Using Supervision

In the previous chapter you were introduced to a particular model for learning from experience. You learned that experience itself coupled with your ability to process that experience meaningfully are key resources in your learning. Chapter 4 also referred to the supervisor's role in helping you process your experiences. This chapter further develops your understanding of the supervisory relationship and how to use it effectively.

A Student's Reflections on the Supervisory Relationship

I think my supervisor is a perfect one for me. I have worked for supervisors before who told me what to do, set up work schedules, and dished out a little praise or criticism as the situation called for. This supervisor is different. She really takes time with me and wants me to learn, but she doesn't hover over me. Sometimes she has thrown me into things that I didn't know how to deal with, and I felt at loose ends. A few times I have even felt annoyed at her for that. As I look back on it, I think she has given me independence and support in about the right amounts. If she had waited until I felt 100% confident to expect me to handle things, I would never have done anything. I think it must be hard for supervisors to find the right balance between expecting enough and expecting too much.

Understanding the Supervisory Relationship

The field supervisor is an important key to student learning in any human services internship. You might think of your supervisor in a number of ways. Your supervisor may be thought of as a teacher who can impart new knowledge and skills. He may be thought of as an enabler, a supportive mentor who can help you to achieve your goals. She might be considered a broker and advocate, an established professional who can help you gain access to key experiences and people within the organization as well as within the community (Brashears, 1995).

Ideally, supervisory relationships are based on trust and mutual understanding. Within the context of a trusting relationship, your supervisor becomes your primary and most immediate source of support while you are on the job (Shulman, 1995). Therefore, supervision should be seen as a valuable, desirable part of your working experience, both in your fieldwork and later in your career. Your willingness to draw upon your supervisor's support, knowledge, and expertise will enhance your learning and reduce your stress throughout your internship.

Despite the supportive nature of most supervisory relationships, many fieldwork students have not experienced professional supervision before and therefore often have mixed feelings about it. The prospect of supervision may be both reassuring and anxiety provoking. Although you want guidance and support, you might also have concerns about having your work scrutinized. You might have fears of not meeting your supervisor's expectations or of your work being criticized.

Most fieldwork students feel more comfortable with supervision once they have formed a relationship with their supervisor. As is the case in other types of relationships, establishing good communication with your supervisor forms a solid foundation for your work together, so it is helpful to discuss with your supervisor your thoughts about supervision. It may also be helpful to recognize that although supervision starts during the human services training program, it is not unique to your role as a student (Neukrug, 1994). Supervision is a working relationship that you will have throughout your career. Wherever you work, whatever you do, you can be relatively sure that you will be accountable to a supervisor. Working effectively with supervisors is a key element in the growth and development of human service professionals throughout their careers (Kaiser, 1997; Sadow, Ryder, Stein, & Geller, 1987).

◆ **EXERCISE 5.1 PERSONAL REFLECTION: OBSERVATION OF SELF AND OTHERS**

Interview a human services professional who has been in the field for some time. Learn what you can from her or him about how to work with a supervisor in a productive way. Potential questions for your interview are listed below, but feel free to generate your own questions based on your interests and situation. Your questions might include:

◆ How have you benefited from supervision?
◆ What challenges have you encountered in your supervisory relationship?

- Have you ever had a serious difference of opinion with your supervisor about how to handle a situation? If so, how did you manage this difference of opinion?
- As I begin my internship and my human service career, what suggestions do you have for me about working with my supervisors successfully?

Kerson (1994) points out that the supervisory relationship progresses through certain predictable developmental stages. She asserts that the relationship's development loosely parallels the eight stages of human development described by Eric Erikson. Hypothetically, following this model, the supervisory relationship begins with the establishment of trust, then moves into the establishment of student autonomy, followed by the development of student initiative, then student industry, and so on. This is a useful way in which to think about the supervisory relationship in that it holds a number of implications. First, it asserts that the relationship develops over time. Second, it posits that the student grows and becomes more independent within this relationship. Third, it implies that the work of supervision is significant as the supervisor's role theoretically parallels that of the nurturing parent and the student's role parallels that of a young child striving to grow.

As this model suggests, learning to work productively with a supervisor and to use supervision effectively is an important professional skill that will help you to continue growing throughout your career. The exercises below will help you clarify your thoughts, feelings, and expectations as you enter this relationship.

Supervisor Characteristics

◆ EXERCISE 5.2 PERSONAL REFLECTION: OBSERVATION OF SELF AND OTHERS

What characteristics do you consider to be ideal in a field supervisor? What hopes and expectations do you have about the type of supervision that you will receive in your internship? What particular concerns or worries do you have about your supervision? What specific requests, if any, would you like to make of your supervisor?

Thomlison et al. (1996, pp. 136–139) identify some traits of "ideal field supervisors." According to this source, the ideal field supervisor

◆ is available,
◆ is knowledgeable,
◆ directs the student's learning, and
◆ has realistic expectations.

Alle-Corliss and Alle-Corliss (1999) suggest that the most effective supervisors are supportive, understanding, open to learning, knowledgeable, experienced, self-aware, and flexible. They further suggest that effective supervisors are able to provide constructive feedback, appropriate relationship boundaries, and mentoring.

Compare these lists of characteristics to your own thoughts about supervision in Exercise 5.2. After examining the listed characteristics, you might have additional thoughts about what you need in a supervisor. It is useful to share your thoughts about supervision with your supervisor. Keep in mind that any requests you might make are just that, requests. Supervisors must decide how to conduct supervision according to their students' learning needs and their own supervisory styles. Sensitive supervisors, however, try to respect your preferences whenever possible, and they can do this best when you have clearly communicated with them about your hopes and needs.

A Student's Reflections on Supervisor Characteristics

I liked my supervisor on sight. She has a wonderful sense of humor, a quick mind, and a ready smile. She has a way of being friendly and businesslike at the same time. We went to lunch together where we discussed both of our goals for the

continued

fieldwork and how we should best go about achieving them. I appreciated her good sense in taking us away from the office as we were getting to know one another. She is a walking encyclopedia of information, but she shares it in a way that is natural and helpful, not superior. I can't help but learn if I spend much time with her.

Although most supervisory relationships develop fairly easily, sometimes there are special barriers to developing a positive relationship. Experiencing your supervisor as overly critical or unavailable to you are obvious examples of this. By the same token, having a supervisor who is far younger than you are or one who has a vastly different background can make forming the relationship more challenging. At times, supervisors and their students simply have very different personalities and personal styles, resulting in distance that may take a longer time to bridge in the relationship.

Whatever the challenges, communication, patience, and flexibility are key ingredients in ultimately building a satisfactory relationship. Although in some cases the relationship may develop slowly, most students and their supervisors find ways to work together in a satisfactory manner. If you should find that your situation is particularly difficult and is creating an obstacle to your learning, you should discuss this with your faculty liaison.

A Student's Reflections on Supervision

My relationship with my supervisor started out a little uncomfortable. He is about 25 years old and has a master's degree and a few years of work experience. Here I am 45 years old, old enough to be his mother. In fact, I do have a son who is 21. I think my supervisor was a little intimidated by me at first. I was quite relieved when he brought up our age difference and asked me how I felt about it. It gave me a chance to tell him that I felt I could learn a lot from him regardless of his age, which seemed to help break the ice. It also made me feel good that he said that he felt he could learn from me, too. I am glad that he doesn't see me like he would a 20-year-old, and I'm glad he had the courage to bring up the subject for discussion. The comfort level between us has been much improved ever since. Now that I am approaching the end of my internship, I can honestly say that I have learned a lot from him. I think the main things that we both brought to our relationship that made it work out so well were honesty, respect, and openness.

Because supervision is a relationship based on two-way communication involving both the supervisor and the student, it is necessary to look at your own contribution to the relationship as well. We now turn our attention to student characteristics and how they influence the supervisory relationship.

Student Characteristics

◆ EXERCISE 5.3 PERSONAL REFLECTION: OBSERVATION OF SELF AND OTHERS

Imagine that you are a field instructor supervising a human service intern. What characteristics would you like to see in the student you are supervising?

How would you like the student to relate to you?

What specific traits and behaviors would you be concerned about if you saw them in your student?

In what ways do you imagine that you would be accountable to others in regards to your student's performance?

Exercise 5.3 gives you the opportunity to empathize with your supervisor and to look at your internship through the eyes of your supervisor and the host organization. Recognizing your supervisor's and your agency's needs in relation to your performance can help you understand the importance placed on the quality of your performance. Taking this empathic point of view will serve you well not only as an intern but also as an employee in the future. Your supervisor has assumed an important responsibility in accepting you as an intern in the organization. Your supervisor will be held accountable by the agency's management for the quality of your performance and therefore has a legitimate interest in knowing what and how you are doing. In this context, it is understandable that your supervisor will want to have a guiding hand in your work and will count on you to seek supervision appropriately.

Although you might feel that your supervisor has all the power and control and that you have none, rest assured that you too have a powerful role to play in shaping the supervisory relationship. The qualities that you bring to the relationship and the manner in which you communicate with and relate to your supervisor are critical to the quality of your relationship and how much you learn from supervision (Halley, Kopp, & Austin,1998). Kaiser points out that supervisees have the power to "enhance or sabotage the supervisory process" (1997, p. 48). Although supervisors have the knowledge and expertise to provide a learning opportunity, the supervisee must be cooperative and open to what the supervisor has to offer.

Through working with many supervisors, it is my impression that certain characteristics are commonly considered to be desirable in interns because they are conducive to a positive learning experience. The "ideal intern" might be described as

- ◆ accessible to the field instructor through openness to feedback and instruction,
- ◆ eager to learn,
- ◆ inquisitive and energetic,
- ◆ knowledgeable on at least a basic level,
- ◆ realistic about his or her own skills and knowledge,
- ◆ willing to take risks in order to gain new skills and knowledge,
- ◆ appropriately assertive, taking responsibility for his or her own learning and demonstrating initiative, and
- ◆ a good listener, observer, and communicator.

This list of characteristics, of course, is not exhaustive. Royse et al. discuss an extensive list of characteristics that are desirable in human services interns. These characteristics include a desire to help others, an interest in the organization, maturity, promptness, reliability, a positive attitude, and good grammar, among others (1996, pp. 16–17). Interestingly, but not surprisingly, both students and supervisors want to work with someone who is accessible and ready to invest energy in the learning process.

The best way to find out about your particular field supervisor's expectations regarding students is to discuss the topic directly. In fact, you and your supervisor probably discussed these expectations at least briefly during your initial meeting prior to the beginning of the internship. At this point, if you do not feel that you are sufficiently clear about your supervisor's expectations of you, initiate some discussion about it. A good starting point for the discussion might be sharing the thoughts you have developed in Exercises 5.2 and 5.3, asking for your supervisor's reaction and input.

Clarifying expectations is an important step in establishing a positive relationship with your field supervisor. The relationship, however, is a means to an end and not an end in itself. As in working with clients, a good working relationship is a necessary, but not sufficient, condition for change and growth to occur. Within the context of this supervisory relationship, a good bit of work must occur throughout the internship.

Working Within the Supervisory Relationship

The process of supervision requires work on the part of the student and the supervisor. You will probably interact with your supervisor in a number of different ways. You will ask a quick question in the hall; she will briefly check in with you on how a particular meeting went. Maybe you will have lunch together at times, and issues related to your work will weave in and out of your lunch-time conversation. You might observe your supervisor's work, and he might observe yours. All of these are valuable parts of your learning experience, but none of them constitute formal supervision.

Formal supervision occurs when you and your supervisor have a *planned contact* with one another, which both of you have *prepared for*, for the express purpose of thoroughly *discussing professional issues*, *planning subsequent interventions or projects*, and *generating feedback* (Alle-Corliss & Alle-Corliss, 1999; Chiaferi & Griffin, 1997; Thomlison et al., 1996; Wilson, 1981). To understand the work of supervision, it is helpful to consider individually each of the key components italicized above.

Supervision Is a Planned Contact

You and your field supervisor should schedule a routine one-to-one meeting time, if at all possible. The norm in most field placements is about one hour per week of formal supervision time. This may vary, of course, depending upon the nature of the work and the issues at hand at any given time. Planned contact ensures that supervision time is a priority and not an activity that occurs "whenever things slow down." In most human services agencies, things almost never slow down. Another benefit of planned contact is that planned meetings allow sufficient opportunity for both you and your supervisor to prepare for supervision.

Supervision Is Prepared For

Your supervisory time is valuable and possibly difficult to arrange. You will want to use the time well. The best method for ensuring this is to prepare for the meeting by developing written summaries of important events that you need to discuss (as discussed in Chapter 4) and an agenda of your concerns and questions. Prior to each supervisory session, your preparation will involve taking a pulse on your current work. Reflect on the work you have been doing, the meetings and other interactions that you have observed, the decisions currently confronting you, the preparations you are making for future contacts with clients or for projects, and the interactions you have had with other staff members. Supervision obviously cannot focus on every experience that you have had and every pending event, so you will have to set some priorities. Questions to ask yourself as you plan for supervision are, "What issues in my work currently concern me most?" and "What are the time-sensitive issues that need my supervisor's attention *now?*"

Once you have decided upon a few priority items for supervision, your preparation continues by gathering the necessary data and organizing your thoughts and questions so that you are ready to present your concerns concisely. Plan your agenda for supervision with the awareness that your supervisor may also be planning to bring in items for discussion. Your supervisor might want to follow a certain project or case closely or might need to use some of the conference time for launching a new task or project in your learning plan. In order to coordinate the concerns that both of you are bringing to the meeting, it is helpful to your supervisor to receive your agenda a day or so in advance. If more time is needed, then this can be scheduled.

Supervision Involves Discussing Professional Issues

As you and your supervisor bring items into the supervisory conference, a twofold purpose is served: (1) the quality of your work for the agency is being monitored and enhanced, and (2) your knowledge, skills, attitudes, values, and behaviors as a developing professional are being encouraged, developed, shaped, and reinforced. Toward these ends, the professional issues discussed may cover a wide range of topics, including but not limited to the following:

◆ your work with clients
◆ projects that you have been assigned
◆ your work and interactions with colleagues
◆ your understanding of and responsibilities to the agency
◆ professional values and ethics
◆ your personal reactions, feelings, attitudes, and biases as these relate to your work
◆ the supervisory relationship itself

As this list suggests, the potential topics for discussion are fairly wide ranging. It is important to clarify, however, that all discussions in supervision are in service to the goals of supervision identified previously—enhancing the quality of your work and your learning. Although supervision can, and often does, touch upon personal issues at times, supervision is not counseling or psychotherapy, nor is it a friendship.

Supervision Includes Planning

Although there are numerous points of departure for discussion in supervision, all of them essentially lead to one overarching question in the end: "What are the implications of this discussion for improving my future work?" Attention is placed on such questions as, "What will I do in my next contact with this client?", "How will I relate to this particular staff member in the future?", "How will I handle myself in the next meeting?", and "How will I approach the next task in completing this project?"

You and your supervisor will examine various options for how to proceed with your work, select an appropriate plan, and discuss its implementation. As discussed in Chapter 4, an essential element in developing any plan for your future work is to consider your ability to implement it. Based on your current level of skill, can you carry out this plan independently? If you do not currently possess these skills, how can you acquire them? In some cases, you might need to implement a given plan with another worker present for support, or you might observe another worker implement the plan as the next step in your learning. Through the process of making and implementing plans, your skills are continuously being reviewed and expanded.

 ◆ **EXERCISE 5.4 SYNTHESIS: LINKING KNOWLEDGE AND EXPERIENCE**

Through this discussion, you have learned about what quality supervision should entail. Examine your own supervision, and consider each of the key points discussed above. To what extent is your supervision:

Planned and systematic:

Prepared for:

Focused on professional issues:

Focused on planning:

Supervision Generates Feedback

Getting feedback is one of the most crucial yet potentially difficult aspects of the supervisory relationship. Emphasis is placed on the issue of feedback in this chapter because your ability to both accept feedback and give it effectively is essential to your success in your fieldwork and in your career.

As a student of human services, you might have learned about giving feedback to others effectively. One of the principles to remember in giving feedback is that feedback can be difficult to receive. Interestingly, receiving feedback of any kind (both positive and negative) can be among the most difficult and challenging tasks of supervision. Yet ongoing evaluation is an indispensable component of your learning and development, not just as an intern but as a professional. Your field experience provides a good opportunity to work on receiving feedback effectively.

◆ EXERCISE 5.5 PERSONAL REFLECTION: OBSERVATION OF SELF AND OTHERS

Think about some specific instances in which you have received positive feedback. (You may draw from experiences within your internship or from other experiences.) In what instance(s) have you been able to accept and benefit from positive feedback? In what instance(s) did you tend to reject the feedback offered? What factors contributed to your acceptance or rejection of the feedback offered?

Think about some instances in which you have received negative feedback. (You may draw from your internship experience or from other experiences.) In what instance(s) have you been able to accept and benefit from negative feedback? In what instance(s) did you reject the feedback offered? What factors contributed to your acceptance or rejection of the feedback?

As you receive feedback, both internal and external factors can influence your ability to accept it. The following internal issues, for example, affect how people receive feedback:

- Feedback is easier to receive when it matches a person's self-perception and is more difficult to receive when it conflicts with a person's self-perception.
- Internalizing feedback requires time and reflection on the part of the receiver.
- Feedback is easier to receive if the receiver trusts the source of the feedback. (Patterson & Welfel, 2000)

A person's ability to accept feedback depends not only on internal factors but on external factors as well. How the feedback is delivered can strongly influence whether that feedback is accepted or rejected. Effective feedback should be:

- direct and specific, describing specific behaviors or actions.
- offered calmly and respectfully.
- timely—that is, delivered soon after the experience.
- balanced, recognizing both strengths and weaknesses.
- offered, not forced, allowing the receiver to reflect and respond.
- helpful in generating ideas for alternative ways of doing things. (Bogo & Vayda, 1998; Egan, 1994; Patterson & Welfel, 2000)

 ### ◆ EXERCISE 5.6 SYNTHESIS: LINKING KNOWLEDGE AND EXPERIENCE

Think about the feedback that you have received during your internship, particularly from your supervisor. In what ways has the feedback that you received generally conformed to and/or violated the guidelines previously suggested?

As your relationship progresses, you and your supervisor might discuss how your supervisory relationship is working. Some supervisors are eager for feedback from their students, asking for suggestions as to how the student's learning needs can best be met. Other supervisors may be less open to suggestions. The relationship works best when the student and the supervisor can give one another feedback about how the supervisory relationship is working. Each of you probably has some suggestions about how you can work together most effectively. If you should have such a conversation with your supervisor, remember that it is important to evaluate your contribution to the success of the supervisory relationship, not just to evaluate your supervisor's role.

Equally important, students must be assertive enough to express any concerns they might have directly to their supervisors. In many cases, interns will complain to their faculty member that their time is not being used well or that the work they are asked to do is not substantive or educational. In response to such concerns, Milnes (2001) asserts, "Approaching one's boss with work-related issues is a reality of employment. The internship is an opportunity to experience that reality" (p. 4). Through taking responsibility for the quality of your own experience, solving problems with your supervisor as they arise, and making the effort to shape the internship into the learning opportunity you want, you will develop important skills that are necessary in any workplace. You will not develop these skills if your faculty member handles these situations on your behalf. Discussing issues of concern with your faculty member, however, is an appropriate way to prepare for a conversation with your supervisor about your concerns.

 ## ◆ EXERCISE 5.7 PERSONAL REFLECTION: OBSERVATION OF SELF AND OTHERS

Using the preceding guidelines regarding giving feedback, write some helpful feedback that you might share with your field supervisor at this point in your relationship, if he or she seems open to such a conversation.

What are the particular strengths that your supervisor brings to your relationship?

How might your supervisor help you to get more out of supervision or more out of your fieldwork in general?

How comfortable are you with the idea of discussing this feedback with your supervisor?

How open do you perceive your supervisor to be to receiving your feedback?

Evaluate your own role as a supervisee. What strengths have you brought to the supervisory relationship?

What have you done to make the most of your supervisory time?

What might you do to get more out of supervision?

Conclusion

Your supervision may be best thought of as a helping relationship in which a professional worker assists you in your learning. Supervision serves the dual purposes of monitoring the quality of your work and enhancing your professional development. For this relationship to work effectively, it must be understood as developing over time and be based on good communication. The supervisory relationship should encompass certain key elements such as planned contacts, preparation, discussion of professional issues, plans for future action, and feedback.

 ◆ **FOR YOUR E-PORTFOLIO**

From your internship, what have you learned about what you want and need in your supervisory relationship? How have you handled your responsibilities within this role in terms of being an active and open participant in the process? What particular challenges have you encountered in your supervision, and how have you handled them? How would you like to grow in your skills to use supervision effectively? Based on your internship supervisory experience, what do you feel you need in a supervisor when you enter your first professional job?

References

Alle-Corliss, L., & Alle-Corliss, R. (1999). *Advanced practice in human service agencies: Issues, trends, and treatment perspectives*. Belmont, CA: Wadsworth.

Alle-Corliss, L., & Alle-Corliss, R. (1998). *Human service agencies: An orientation to fieldwork*. Pacific Grove, CA: Brooks/Cole.

Bogo, M., & Vayda, E. (1998). *The practice of field instruction in social work: Theory and process*. New York: Columbia University Press.

Brashears, F. (1995). Supervision as social work practice: A reconceptualization. *Social Work, 40*, 692–699.

Chiaferi, R., & Griffin, M. (1997). *Developing fieldwork skills: A guide for human services, counseling, and social work students*. Pacific Grove, CA: Brooks/Cole.

Egan, G. (1994). *The skilled helper: A problem management approach to helping*. Pacific Grove, CA: Brooks/Cole.

Halley, A., Kopp, J., & Austin, M. (1998). *Delivering human services: A learning approach to practice*. New York: Longman.

Kaiser, T. (1997). *Supervisory relationships: Exploring the human element*. Pacific Grove, CA: Brooks/Cole.

Kerson, T. (1994). Field instruction in social work settings: A framework for teaching. In T. Kerson (Ed.), *Field instruction in social work settings* (pp. 1–32). New York: Haworth.

Milnes, J. (2001). Managing problematic supervision in internships. *NSEE Quarterly, 26*(4), 1, 4–6.

Neukrug, E. (1994). *Theory, practice, and trends in human services: An overview of an emerging profession*. Pacific Grove, CA: Brooks/Cole.

Patterson, L., & Welfel, E. (2000). *The counseling process* (5th ed.). Stamford, CT: Thomson.

Royse, D., Dhooper, S., & Rompf, E. (1996). *Field instruction: A guide for social work students* (2nd ed.). White Plains, NY: Longman.

Sadow, D., Ryder, M., Stein, J., & Geller, M. (1987). Supervision of mental health students in the context of an educational milieu. *Human Services Education, 8*(2), 29–36.

Shulman, L. (1995). Supervision and consultation. In R. Edwards & J. Hopps (Eds.), *Encyclopedia of social work* (pp. 2373–2379). Washington, DC: NASW Press.

Simon, E. (1999). Field practicum: Standards, criteria, supervision, and evaluation. In H. Harris & D. Maloney (Eds.), *Human services: Contemporary issues and trends* (pp. 79–96). Boston: Allyn & Bacon.

Thomlison, B., Rogers, G., Collins, D., & Grinnell, R. (1996). *The social work practicum: An access guide*. Itasca, IL: Peacock.

Wilson, S. (1981). *Field instruction: Techniques for supervisors*. New York: The Free Press.

Young, M. (1998). *Learning the art of helping: Building blocks and techniques*. Upper Saddle River, NJ: Prentice-Hall.

Chapter 6

Communicating with Clients

One of the most strongly emphasized components of human services education is communication skills. In your internship, you have the ideal opportunity to practice and strengthen these skills. Although the human services curriculum tends to parcel out this skill development into discrete courses about working with individuals, families, groups, and communities, the typical workday in a human services organization is not so compartmentalized. As an intern, you may find yourself quickly shifting from working with a family to meeting with a community task force to accepting a phone call from a distressed individual whom you will see later that afternoon in group. Professionals who have mastered the communication skills necessary to work successfully with each of these levels of intervention are able to shift relatively seamlessly and effortlessly from one point of focus to another. Your internship gives you the opportunity to practice this type of mastery in your skills.

Although this task no doubt seems challenging, the good news is that the same basic communication skills comprise the essential components of effective work with individuals, families, groups, and communities. Each level of intervention, however, does require a number of additional skills in the helper. In this chapter, we first turn our attention to the core communication skills and then examine specific issues involved in each level of intervention: individual, family, group, and community.

The Basic Skills

Although it may be common sense to think about communicating as the process of verbal and nonverbal interaction with another person, this is not the best place to start in developing your communication skills. Both verbal and nonverbal communication spring from who the communicator is at a deeper level, reflecting the communicator's values and attitudes. Therefore, as you begin to think about your communication in your internship, it will be beneficial to think first about the values, beliefs, and attitudes that you observe in yourself in relation to your work in the field.

Values and Attitudes

The human service literature identifies the specific values and attitudes that lend themselves to worker effectiveness as well as those that interfere with or block worker effectiveness. The desirable worker values and attitudes are well-known to students in human services. They include empathy, nonjudgmental attitude, acceptance, genuineness, respect, warmth, patience, open-mindedness, rejection of stereotypes, and a positive view of human nature (Mandell & Schram, 2006; Neukrug, 2004; Okun, 2002; Woodside & McClam, 2006). Conversely, we know that the values and attitudes that negatively affect professional communication include stereotyping, judging, rejecting attitudes, closed-mindedness, phoniness, impatience, and negative views about the human capacity for change.

Because human services professionals are human, it is to be expected that at times our attitudes and reactions toward the people and situations we encounter may be less than ideal. Although we cannot expect to reach the ideal of perfection in our values and attitudes, it is important that we cultivate self-awareness, a self-observing process that alerts us when we are experiencing reactions within ourselves that make positive, constructive communication difficult. It is not constructive to be unrealistic with ourselves about negative reactions that we feel. If we are having these reactions, they do not just go away if ignored. Instead, they tend to creep out in our communication in harmful ways that we may not even be aware of. Awareness is the first step in dealing with these reactions responsibly and professionally.

◆ EXERCISE 6.1 PERSONAL REFLECTION: OBSERVATION OF SELF AND OTHERS

What reactions do you observe in yourself at your internship?

Reflect on how you feel when interacting with or observing clients and staff. In what situations and toward what people have you experienced the desirable reactions noted above?

In what situations and toward what people have you experienced the less desirable reactions?

Ideally, you become aware of counterproductive reactions in yourself before they have had an opportunity to affect your communication negatively. Once these reactions have surfaced, there are a number of steps that you can take to process them and put them in perspective. Simply reflecting on where these feelings are coming from can be constructive. Understanding how your reaction is stemming from your own personal history may help you put your reaction aside, seeing it as somewhat irrelevant to the present situation. When you are able to "own" your feelings as being more about you and your history than as an indicator that something is wrong with a client or co-worker, you will have made a giant step in laying the groundwork for positive communication. Talking with someone else about your reactions can also help you to ventilate those feelings and then put them aside so that they will be less likely to harm your work. Talking with your field supervisor, your faculty supervisor, a staff member, or a class-mate provides a good forum for processing your reactions as well as possibly getting some helpful feedback from someone you trust. After having dealt with their reactions responsibly, most professionals are then able to put them into their proper perspective and engage in productive helping interactions. If you find that you are encountering persistent reactions in yourself that limit your effectiveness, you should discuss this with your supervisor so that decisions can be made about how to handle the situation.

◆ **EXERCISE 6.2 PERSONAL REFLECTION: OBSERVATION OF SELF AND OTHERS**

Working in a human service organization as an intern, you probably have a wonderful opportunity to observe the communication of many staff members throughout the day.

Who do you find to be a good communicator? What aspects of this person's communication do you admire?

Is there anyone who, in your opinion, needs to strengthen their communication skills? What particular aspects of this person's communication need to be modified? Values and attitudes? Nonverbal communication? Verbal communication?

How do you think others in the organization might describe you as a communicator? What strengths and weaknesses might your co-workers attribute to you?

Active Listening

The most fundamental communication skills for human service professionals are best reflected in the verbal and nonverbal aspects of active listening. Active listening consists of a set of skills that will serve you well, not only within your professional roles but also within all social contexts. Practicing the skills of good active listening throughout your internship not only enhances your skill development, it also enhances the quality of your internship through better communication with your field supervisor, faculty supervisor, staff members, and clients.

Active listening in its nonverbal aspects calls upon the communicator to be fully engaged and focused, concentrating on the speaker. Facing the speaker squarely, maintaining eye contact, maintaining a relaxed, open posture (refraining from crossing your arms or legs), leaning slightly toward the speaker, and nodding are basic components of nonverbal active listening.

The verbal aspects of active listening include such responses as minimal verbal response (e.g., "mm-hmm" and "I see"), paraphrasing or restating the speaker's message, reflecting the speaker's feelings, and occasionally asking an interested question that encourages the speaker to elaborate on a given topic or theme.

As is probably obvious from this description and the name itself, active listening is primarily about being an attentive, focused listener. Responses, both verbal and nonverbal, then flow from what you are hearing as you engage in focused concentration on the other person (Mandell & Schram, 2006; Okun, 2002).

◆ EXERCISE 6.3 SYNTHESIS: LINKING KNOWLEDGE AND EXPERIENCE

What factors might hinder your ability to be an active listener?

How might these challenges be overcome?

In your internship, how do you see the skills of active listening employed?

When do you see active listening being used by staff?

To what extent are you practicing the skills of active listening in your work?

Think of at least one conversation in which you employed active listening and evaluate your skills. What specific goals might you set for yourself as you seek to strengthen your active listening skills in your internship?

Throughout your internship, employ your active listening skills in working with clients. In doing so, you will build upon these basic skills to conduct thorough helping interviews in which clients' concerns are more fully explored and decisions are made about specific actions that might be taken to address those concerns. In the following section, we examine more closely the skills involved in client interviewing as well as specific exploration skills and action skills employed by human service professionals.

Client Interviewing

There may be a risk in referring to a helping conversation as a helping interview because to many the term *interview* implies a question and answer session. This is not the case with helping interviews. In the context of human service delivery, an interview is simply a conversation in which a professional guides the interaction in such a way that there is a beginning, middle, and end, with specific tasks being completed within each stage (Brill, 1995).

There is a beginning in which initial rapport is established. Introductions are made, some very brief small talk might ensue, and the worker initiates discussion about the client's concerns. The client's concerns are then briefly explored to the extent that both the client and worker can be satisfied that the client is at the right place speaking with a person who can potentially help in some way. This is sometimes referred to as "establishing a contract." This contract may be somewhat formal in some settings with clients signing intake papers, agreements, etc. In most settings, this "contract" is fairly informal as the worker and client mutually agree (often tacitly) to engage in the helping process more fully.

The conversation then moves to the middle stage in which there is a more thorough exploration of the client's concerns, followed by consideration of how those concerns might be addressed. In some cases, there might be immediate information and assistance that the worker can provide. In other cases, there may be aspects of the situation that will require more sustained efforts, conversations, interventions, and/or referrals.

The task of the middle stage is to ensure that an adequate understanding of the client's situation and concerns has been developed and that preliminary plans for assistance are underway. It can be understandably frustrating for a client to leave a helping encounter feeling that they have "just talked." Although it is not realistic to expect that the professional will have all the answers or solutions, the worker should take responsibility for helping the client develop a plan focused on "where to go from here." Such a plan builds the expectation for positive change, infuses the process with a sense of hope and direction, and provides momentum for the actions that need to be taken in order for issues to be resolved.

As the conversation comes to an end, there are specific elements that the professional should ensure are included in the closing of the interview. There should be a summary of the important points of the interview in which the worker conveys an understanding of the client's concerns and reiterates the plan. New material should not be introduced during the closing. Instead, it is time for tying up loose ends and clarifying future directions. The worker should assume responsibility for writing down any specific commitments either party has made, clarifying that the client understands next steps, ensuring that the client has the tools needed to make those steps (i.e., phone numbers, names, information, etc.). Sufficient time should be allowed for the closing so that this stage is not rushed or confusing. Conducting the end of the interview well is quite important in establishing follow-through and future contact with the client.

It is certainly the case that not all human service settings provide the kinds of services that lend themselves exclusively to such formal helping conversations. Nevertheless, the same general principles can and should be applied when working in these less formal contexts. Working in a recreation center, afterschool program, group home, residential treatment program, retirement center, or other such settings will give you more opportunities to communicate with clients over the course of the day, sometimes around very routine activities and daily tasks and other times around more significant challenges, life events, and emotional issues.

The professional in this context applies active listening throughout the day and shifts into a more in-depth helping conversation with clients as various situations dictate. The retirement home resident who is grieving the move from her own home may at one moment be chatting with you superficially about the furnishings in her room and in the next moment choose to share with you her feelings and fears, asking for your help. The child on the playground at one moment might be playing a friendly game of basketball and in the next moment talking with you about his feelings that the other children leave him out and don't like him. At such moments, you might enter a more in-depth helping conversation similar to that described above, drawing upon its stages and principles to guide your work.

 ## ◆ EXERCISE 6.4 SYNTHESIS: LINKING KNOWLEDGE AND EXPERIENCE

In what situations do you see yourself and staff members engaging in client interviewing as described above?

In what ways do you see beginnings, middles, and ends in your conversations with clients?

Think of a particular helping conversation that you have had and reflect upon that conversation, trying to identify the various stages. As you continue your work, what particular aspects of client interviewing would you like to strengthen?

Exploration Skills

As is probably obvious from the description of the helping interview, active listening alone may not be sufficient to take you through the entire helping process. In fact, it is likely that additional skills will be needed in order to move through the various stages of the helping process productively. Specific exploration skills will help you explore and understand the client's concerns more thoroughly. Some of these skills are summarized below:

Probing: questions or statements that help clients tell their stories or delve more deeply into an issue, generally are best stated as requests for information such as, "I'd like to hear about your family" or "I'm interested in what led you to leave your last job" (Egan, 2002; Egan, 2006; Meier & Davis, 2005; Neukrug, 2002; Okun, 2002).

Clarification: statements or questions that attempt to clear up confusion or to check the client's perceptions, such as, "What do you mean by 'dishonest?'" or "It isn't clear to me whether you are saying that this situation was entirely your teacher's fault or whether you're saying that you had a hand in it as well" (Capuzzi & Gross, 2007; Egan, 2002; Okun, 2002).

Interpreting: statements that might add to the client's understanding of a situation, such as identifying underlying assumptions, feelings, or possible cause and effect, "When I hear you talk about your classmates, you seem to assume that they don't want you around" or "I get the sense that you feel guilty and down on yourself when you have to ask for help" (Neukrug, 2002; Patterson & Welfel, 2000; Welfel & Patterson, 2005).

Confronting: statements that point out discrepancies that clients may need to be aware of and deal with in order for their concerns to be resolved, such as, "You say you're not sad about this, but it looks like you are trying very hard not to cry" or "You seem to see yourself as a very conscientious student, yet you tell me that you've turned in an assignment late and have missed a number of classes this semester" (Neukrug, 2002; Okun, 2002; Patterson & Welfel, 2000; Welfel & Patterson, 2005).

Summarizing: statements in which the worker draws together and paraphrases the major themes and issues presented by clients. Clients can then fill in gaps in the information or clarify points, if needed. For example, the worker might state, "As I understand it, you are needing immediate help to pay your bills so you won't be evicted or have your electricity turned off. But you also want help finding a better paying job or getting some other types of assistance so you won't be in this position month after month" (Egan, 2002; Egan, 2006; Meier & Davis, 2005; Okun, 2002).

The above exploration skills, coupled with active listening, will help you move the helping conversation toward greater clarity and depth of understanding of the client's concerns.

Action Skills

Many of these same skills can also be used as "action skills" as you help clients identify strategies for resolving their concerns. Additionally, skills such as informing, suggesting, and referring may also be used to help a client develop a plan.

Informing: Tapping your knowledge and sharing it with clients in order to empower them as they deal with their concerns (Egan, 2002; Neukrug, 2002, Okun, 2002). For example, "Studies have shown that most children with ADHD are more successful when they not only take the prescribed medication but also have a behavior management plan at home. I can refer you to someone who can help you develop a behavior management plan if you would like." Or, "You might qualify for food stamps or other financial assistance, but I will need to gather some information about your income and expenses to check that out."

Suggesting: Putting forth an idea that you as a professional think may offer a good alternative. It is important that the idea is offered, not pushed, and that it is based on solid information and knowledge, not mere opinion. For example, "Since you have already tried many different outpatient treatments for your son, you might consider some inpatient options. Inpatient treatment can offer more intensive therapy as well as around-the-clock supervision, which might be very useful at this point" (Egan, 2002).

Referring: Facilitating clients' connection with another resource that is needed in order to address their concern. This is more than just mentioning an agency name or writing down a phone number. For example, "I think it is very important that you call Family Abuse Services very soon. Here is their phone number and the name of the crisis worker that you will probably speak with. Please feel free to use my phone and call now if you would like." If the client does not make the contact in your presence, then following up in a few days is generally a good idea to ensure that the contact was made. Also exploring with clients their feelings about requesting services from a particular organization or for a particular problem can help identify and minimize any barriers to the referral that might be present (Brill, 1995; Woodside & McClam, 2006).

 ◆ **EXERCISE 6.5 SYNTHESIS: LINKING KNOWLEDGE AND EXPERIENCE**

As you think about the exploration skills and action skills above, identify specific encounters in which you have used these or have seen a staff member use these skills. Cite a few examples below.

Which of these skills do you feel that you most need to strengthen in yourself?

How can you go about strengthening these skills?

A Student's Reflections

My internship puts me in contact with some of the most difficult people I have ever worked with. Almost all of the kids in the program are diagnosed with oppositional disorders or conduct disorders. Several of them have histories with drugs and arrests for crimes like breaking and entering or assault. It is really hard not to be judgmental at times and I vacillate between feeling angry and scared when they get bent out of shape about something. I see the staff being really strong, staying calm, setting limits, giving consequences. At times they have to use physical restraint to control residents, which is something I'm not trained to do. Being in this internship has shown me how much personal strength it takes to do this work. I have some good role models here, and I'm working on taking a more active role in the treatment program. I know I have to establish my own place and authority with these kids and nobody else can do it for me. Every new staff member gets tested. My supervisor has been great in letting me know that it is normal to feel intimidated at first, but he also expects me to eventually take more of a lead in communicating with the kids, especially when they aren't doing well. The mantra here seems to be communicating caring while setting clear limits and giving consequences. I'm going to learn a lot here, but it's not going to be easy.

As you work in your internship, your host organization may provide you with opportunities to interact with individuals, families, groups, and/or communities. In doing so, you will find that other issues beyond the application of discrete communication skills may come into play in your communication. We now turn our attention to some of the issues that may impact communication within each of these levels of intervention.

Working with Individuals

Much of the literature in human services seems to assume or imply that most if not all worker–client communication fits into the broad category of counseling. Furthermore, there often seems to be an assumption that all clients are motivated and cooperative. After even a very brief time in your internship, you have probably found that many clients are reluctant clients who would not be seeking help if they had a choice. Additionally, the professional's role with and on behalf of clients is generally multifaceted, going well beyond the role of counseling. Human service professionals today engage extensively in case management and advocacy with their clients, roles that often require additional types of communication skills. In this section, we examine the communication skills involved in working with reluctant clients as well as those required for effective case management and advocacy.

Working with Reluctant Clients

Working with clients who are hostile or uncooperative can be very challenging for most human service professionals. Reluctant clients often have been coerced to seek help due to a court order or the expectations of authority figures in their lives. At other times, they might have sought help voluntarily but in the process encounter ambivalent feelings about engaging in or proceeding with the process. Whatever the case, there are frequently times when workers encounter clients who seem unwilling to engage in the helping process. To a certain extent, the professional should continue to use all of the same procedures and skills that have already been discussed in this chapter. However, there are also some adaptations of these skills that are generally recommended when working with reluctant clients.

The themes that should be especially stressed in your own attitudes and disposition as you work with the reluctant client include respect and empathy for the client's perspective. It is easy for most of us to understand the resistance created by being forced to do something we do not want to do. By accessing this genuine understanding in ourselves, we can feel and express empathy for the client's situation. It is also easy for most of us to understand how difficult it is to ask for and accept help, even when we want and need it. Again, through our own experience we can develop true empathy for the ambivalence that even a more motivated client might feel in the helping process. This empathy and understanding are among the most important components of working successfully with reluctant clients.

Also important is the basic principle of all helping relationships—starting where the client is. Talking with clients about their reluctance gives them permission to express their feelings verbally rather than simply to act them out through silent, hostile, or oppositional behaviors that are barriers to developing a relationship. This is not to suggest that talking about those feelings will magically make the feelings go away or that it will ensure that the client will not be hostile or passive-aggressive. Your talking with clients about their feelings of reluctance demonstrates your acceptance of where they are and that you are not frightened or offended by those feelings. It is important as you engage in these conversations that you not try to talk clients out of their reluctant or hostile feelings. Doing so might set up a more polarized conflict or heighten clients' resistance because they feel compelled to counter your arguments or position. In most cases, simple acceptance is more productive in the long run.

The difficulty presented by reluctant clients becomes more difficult if their behaviors trigger strong, negative emotions in the worker. Reluctance in clients can be very threatening to professionals and may trigger feelings of powerlessness, anger, resentment, or the urge to control. All of these reactions are counterproductive. They create more distance between the worker and the client and tend to increase client reluctance. One of the major challenges in working with reluctant clients, in fact, is not allowing yourself to be emotionally reactive to their resistance in these ways.

In broaching the work of the interview itself, you are wise to use less directive methods with reluctant clients. Using active listening methods and giving clients the opportunity to tell their own story in their own way helps them feel less threatened. Their feelings of safety can increase as they feel an increased sense of control.

Questioning reluctant clients generally is not the best approach. When more direct probing is needed, using stated requests for information rather than direct questions is generally more fruitful. Work with reluctant clients proceeds more slowly than with more motivated clients, so you should set reasonable expectations for yourself as the helper. Do not expect to "win the client over" or to create dramatic change in a short time. Confrontation and other more challenging aspects of the helping process should not be rushed. Allow the relationship time to develop (Okun, 2002; Welfel & Patterson, 2005).

Compounding an already challenging situation is the fact that many reluctant clients are often involuntary clients. These are clients who have been, or feel they have been, coerced into seeking services often because their behaviors have been perceived as violating social norms and laws (Rooney, 1992). Parents who have been charged with child abuse and/or neglect, partners who have been abusive to their mates, young people with severe conduct disorders, and adults with severe antisocial behavior are in this group. Their behaviors in themselves generally conflict strongly with the values of the human service professional and can create yet another barrier to establishing a productive working relationship. If you find that you are working with a client whose behaviors evoke strong negative or rejecting feelings in you, remember the importance of dealing with these feelings appropriately. Talking with a supervisor, faculty member, or staff member about your feelings will help ensure that you are handling these emotions in a productive manner.

◆ EXERCISE 6.6 PERSONAL REFLECTION: OBSERVATION OF SELF AND OTHERS

What exposure have you had to reluctant clients in your internship?

What has been your emotional reaction to such clients?

As you think about interactions with reluctant clients that you have been a part of (either as a participant or as an observer), what worker behaviors did you find helpful or unhelpful in the situation?

What suggestions do you have for specific ways that these interactions might have been handled more productively?

Through practice with and exposure to reluctant clients, you will find that your comfort in working with them increases. Some workers eventually even report that they enjoy the challenge of working with less motivated clients and often feel especially empathic toward them and their situations. Once the relationship is established and the worker learns more about the client's personal story, it is often the case that a deeper understanding of the client's situation and personality evolves. As a result, in many cases the worker may develop great respect for the client's strengths in having dealt with significant challenges.

Case Management and Advocacy

In addition to providing direct services to clients, many human services professionals today also provide case management for the clients they serve. As a case manager, the professional coordinates multiple services for the client to ensure quality and continuity of care. This role has become increasingly important in today's human services environment in which resources are often relatively scarce and community services are highly fragmented.

Clients with multiple needs and/or chronic conditions inevitably must rely on an array of varied services provided by numerous agencies and organizations. Case management services are essential in these situations to make appropriate referrals and to ensure follow-through on those referrals. Case managers often must assume an advocacy role for clients to help them make the needed connections with service providers and to prevent them from dropping through the cracks due to bureaucratic procedures, overworked providers, misunderstandings, or other such barriers (Rose, 1992; Vourlekis & Greene, 1992; Woodside & McClam, 2006).

In the case management role, human service professionals' communication skills must enable them to work with clients to identify their needs as well as to work with community professionals to plan and implement client care in a systematic, cooperative manner. Furthermore, the worker must have the skills and values necessary to ensure client involvement, empowerment, and self-determination throughout the process. The communication skills utilized by the case manager are to a great extent the same active listening and action skills described earlier in this chapter. Some additional skills may be needed as well in order to work effectively with the community.

Case management works best when professionals who are working together to assist clients know and trust one another. For this reason, skills in listening and relationship building will be helpful in working with colleagues in the community just as they are helpful in working with clients. Getting to know the missions of various organizations in your community, the services they provide, and the parameters within which these services are delivered is important to making appropriate referrals and to developing realistic expectations about what particular community agencies can do for your clients. The more you know about each agency's resources and constraints, the better you can determine whether your client is receiving appropriate services or whether it is appropriate to ask for more in either the quality or the quantity of those services.

Working in a respectful, collaborative manner with colleagues is always more advantageous and more professional than working in an adversarial, conflictual manner. When conflicts do arise, they are best addressed by striving to understand the perspective of each participant and seeking common ground. Assuming that all parties involved want to do what is best for the client and framing issues in this positive manner help foster cooperation and minimize conflict while keeping conversations focused on the task at hand. When conflicts do arise, case managers must use conflict resolution strategies in order to move beyond the conflict on behalf of the client. Conflict resolution is discussed in some detail in Chapter 10, Managing Your Feelings and Your Stress.

To the extent possible, case managers should educate clients about the services that might meet their needs and empower clients to select and pursue those services. It is important that professionals not do for clients what clients can do for themselves.

The ultimate goal is for clients to be self-directed, not needing professional assistance. Nevertheless, because human service professionals have more knowledge of the human services system and established relationships with agency personnel, they are often better equipped to secure services than is the typical client. In such cases, the principle of client empowerment requires that we "do with" rather "do for" clients.

Case managers should see their clients as capable and treat the relationship with the client as a partnership in which the two are working together toward getting the client's needs met. Even in best case scenarios in which clients are able to do much of their own work in securing services, a professional case manager is often needed to ensure coordination of care when multiple service providers are involved. Professionals planning together can ensure that all services are working toward common goals, that there are no critical gaps in services, and that there is no duplication of services.

Once a service plan has been constructed for a client, the case manager must monitor the plan's implementation by staying in touch with the client as well as the service providers. Monitoring is often most effectively accomplished through periodic "team meetings" attended by the client, the case manager, and the various service providers. The case manager facilitates such meetings, ensuring that the plan is being implemented and that the services provided are accomplishing the intended goals. Routine monitoring also provides opportunities to identify new client needs that may emerge and to alter the plan accordingly. Additionally, such upheavals as changes in agency services and turnover in service providers are less disruptive to clients if they have been a part of team meetings where such changes have been anticipated and planned for. As is obvious from this description, the case manager often functions as a team leader and service coordinator. In this role, skills as a discussion leader and facilitator are essential.

As with all communication skills, becoming a good discussion leader and facilitator takes practice. However, some general guidelines for leading case management discussions are offered here:

1. Develop a clear set of desired outcomes for the meeting and state them at the beginning of the discussion for focus and direction.
2. Raise questions and encourage brainstorming of alternatives.
3. Engage each and every person in the discussion.
4. Listen carefully and practice active listening to ensure that messages are being delivered and heard accurately.
5. Engage group members with one another, not just with you.
6. Engage the group in decision making as appropriate by exploring and weighing various options.
7. Stay out of the middle of conflicts or differences in perspectives while striving to resolve them by clarifying communication and finding common ground.
8. Try to ensure that people are hearing what is really being said as opposed to hearing what they want to hear or hearing what they fear they will hear.
9. Summarize and clarify decisions verbally, recording important information in writing. This may be done on a flipchart or white board to help the group focus visually on the task at hand. A written summary should be sent out to group members to clarify the plan and who agreed to be responsible for what portions of the plan.

10. Acknowledge everyone's contribution to the group's work.
11. Keep clients central to the process by allowing them to make decisions as much as possible and by asking for their input throughout the discussion.

Despite the emphasis on client empowerment, many clients are often not in position to be effective advocates for themselves. This may be the case due to illness, disability, lack of critical resources such as a telephone or transportation, lack of understanding of the human service systems, or the fact that the client is a minor child. In such instances, the case manager may have to assume a far more active role in securing the needed services and may become a more direct advocate for the client. As an advocate, the professional seeks to do what clients cannot do for themselves. In such cases, the worker will more assertively pursue services, not simply provide a referral. An assertive client advocate articulates the clients' needs compellingly and is persistent in addressing services that are lacking in either quality or quantity for the clients they serve (Ezell, 2001). Good assertiveness skills (as discussed in Chapter 10, Managing Your Feelings and Your Stress), in conjunction with the other skills discussed in this chapter, are fundamental to effective advocacy.

 ◆ **EXERCISE 6.7 PERSONAL REFLECTION:
OBSERVATION OF SELF AND OTHERS**

In what ways do staff members in your organization provide case management services to clients?

What case management tasks do you perform in your role as an intern?

Have you observed any meetings or conversations in which staff members have collaborated with professionals from other agencies around case management issues? If so, what did you learn from your observations?

Working with Families

Most human service settings do not focus exclusively on individuals. Organizations that provide services for individuals often focus on their families as well because of its significant role in the lives of most individuals. Programs for children tend to have a particularly strong emphasis on families because children are dependent upon their families for physical, psychological, and emotional support. Many organizations have families as their primary focus through such programs as parent education, family counseling, foster care or adoption services, child abuse or neglect intervention, family abuse crisis intervention, and support of family caregivers. Also, schools and other educational settings increasingly provide programs for parents and families because they recognize that children are more successful academically when their parents are closely involved with the child's education.

Although the basic skills in interviewing, active listening, exploring, and action (discussed earlier in this chapter) apply when working with families, there are also unique issues in communicating with families that come into play in your work. Family work can be particularly challenging. Family members present with diverse ages, developmental levels, experiences, perspectives, and needs. The divergent views represented within a family must be heard and affirmed while also moving toward common goals. Simply forming a relationship with each person in the family and gaining each family member's trust can be challenging in itself. Family interviews can generate enormous amounts of information that can be difficult for the professional to attend to, observe, and respond to. While this discussion cannot present a comprehensive review of the skills required in working with families, it is useful to review a few of the major guidelines in working with families that you are likely to apply in your internship.

Human service professionals frequently conduct family interviews. In this discussion of family interviewing, the assumption is made that two or more family members are present. Family interviews may consist of all family members, all members of a household, or any subsets of these groups. Parent–child dyads, married or unmarried couples, sibling groups, or any other combinations of family members may be present. Extended family members may be present as well as non-kin or neighbors outside the household who function as family members. Family-centered practice suggests that the family determines who is in the family and who should be present for a family intervention (Petr, 1998). Some agencies make a strong effort to have the whole family present for intake interviews and other interventions. This is especially common in mental health agencies or other settings in which in-depth family counseling is delivered.

Students who are just beginning to work with families can often feel anxious and ill-prepared for the challenges they present. Keeping in mind some general principles to follow in working with families can help you approach the family interview with greater confidence.

As you begin the family interview, it is important to make contact with each family member individually. Introducing yourself to each family member and asking each person a little bit about himself or herself conveys your interest in each person and begins to build rapport. Respond briefly to each person in an affirming manner, drawing upon your skills in active listening (Hanna & Brown, 2004). This process is referred to as *joining* (Minuchin, 1974). This small-talk stage should be brief with each person so that it does not drag on. Brevity is especially important with larger family groups.

As the interview continues, it is important for you to observe and respect the family's patterns of interaction and to respect the hierarchy within the family. Engaging with a family in a manner that respects its internal organizational structure is referred to as *accommodating* (Minuchin, 1974). For example, in most families, the parent(s) holds a position of authority and family leadership. Therefore, it is appropriate for you to introduce yourself first to the parent(s) and to seek their input first about the family's concerns. Out of respect for this hierarchy, it is also best that the worker not be critical of parents while the children are in the room. If you need to give such feedback to parents, it is best to arrange a separate time without the children present so the parents do not feel that you are undermining their authority.

On the other hand, while being respectful of the family's hierarchy, it is important that you do not allow yourself to be controlled by the family's counter-productive patterns of interaction. For example, as the professional, you will want to gain the perspective of each family member on various issues. In some families, the egalitarianism that this demonstrates may be quite contradictory to their routine ways of functioning. By continuously making contact with all members of the family, you not only gain more complete information but you also demonstrate that everyone's perspective matters and that everyone in the family has a role in improving the situation at hand (Minuchin, 1974).

One commonly used interviewing method in family work is circular questioning (Thomlison, 2002; Young & Long, 1998). Circular questioning is a method of eliciting the family members' perspectives about events and relationships, particularly focusing on periods in which there have been significant shifts in family functioning. The depiction of this type of questioning as "circular" refers to its aim of identifying recurrent patterns of interaction within the family as family members comment on their perceptions of one another's behavior in varied situations. Examples of circular questions are "Who worries more about your daughter?" "Who noticed the problem first?" "How would things be different if he did this instead?" "Who agrees with you that this is what happens?" "Who did what then?"

Through such questions the family can begin to recognize cycles of interaction that are predictable and can come to recognize the circular causality of problems. From these insights, the family can recognize various points in the cycle where change might occur. As this understanding of family interaction grows, the causes of certain behaviors are no longer understood as simply linear (A causes B) but as circular (A causes B, which causes C, which causes A). Most family experts agree that circular causality and reciprocal causality (A causes B and B causes A) more accurately and adequately describe family functioning than does simple linear causality (Goldenberg & Goldenberg, 2000; Thomlison, 2002; Young & Long, 1998). Furthermore, the primary purpose of circular questioning is to produce change rather than to uncover the "truth."

In working with families, it is important that you look for, recognize, and build upon the family's strengths (Ragg, 2006; Thomlison, 2002). In the past, many human service professionals have tended to blame families for problems and to see the family primarily as a source of pathology. Families seeking help are often sensitive to this issue. The family is strengthened by seeing itself as resourceful. Pointing out to families their strengths, instances in which they have coped well with situations, displayed resilience in dealing with difficulty, and shown the care and concern they have for one another, can go a long way in producing a more positive family identity and a more positive relationship with you as the helper. It is important that such comments rest on genuine observations of family strengths rather than facile, unfounded compliments. Otherwise, your efforts will most likely be counterproductive.

An additional challenge in working with families is maintaining neutrality, assuming a nonjudgmental point of view toward all parties (Young & Long, 1998). As the professional, you must strive not to be co-opted into an alliance with one particular person or subgroup within the family. You must maintain the flexibility to empathize with various points of view and support various family members at different times. Maintaining this neutrality requires a high degree of self-awareness on your part and reflection on your own family dynamics and interaction patterns.

In the absence of self-awareness, it is not uncommon for professionals to be unduly influenced by their own family experiences. It is often easy, for example, as the youngest child in your family to overidentify with the youngest child in the families you are working with. Similarly, it is tempting to expect families to interact as your family did and resolve problems in similar ways.

As a neutral helper, you must have the flexibility to work with each family as unique and capable of choosing its own ways of interacting and solving problems. That said, one important caveat regarding neutrality is essential to address. Neutrality does not mean that as a helper you cannot confront and take a stand against abusive behavior. As a professional, you must intervene as needed in an effort to stop abusive behaviors and to address appropriately concerns such as substance abuse and mental illness (Young & Long, 1998).

Families that include children often present circumstances in which the helper observes behavior in the child that needs to be corrected in some way. When the child's parent is in the room, the helper should generally defer to the parent to set limits on the child's behavior (Petr, 1998). Simply observing the parent's method and timing in dealing with inappropriate behavior provides important assessment information while also recognizing the parent's authority. If you are growing increasingly uncomfortable with a child's behavior, you might ask the parent to intervene. For example, a child who is repeatedly kicking the wall may raise your concern that the wall will be damaged or the child may get hurt. In such a situation, you might ask the parents to intervene. At this point, they might do so or let you know that it is acceptable for you to do so. In either case, you will not have entered the parents' domain of authority uninvited. Should you be invited to intervene, this presents an opportunity for you to model appropriate limit setting.

◆ EXERCISE 6.8 SYNTHESIS: LINKING KNOWLEDGE AND EXPERIENCE

In what situations do you have the opportunity to work with or observe families?

In what instances might you be able to apply the skills and practices described above in your work?

Working with families can be especially rewarding because producing positive change in a family has the potential to improve the quality of life for several people. Similar benefits are also applicable to working with groups because the impact of the work you do in a group can be experienced by many.

A Student's Reflections on Working with Families

Working with a social worker in a hospital, I see lots of families every day and interact with families that are making decisions about taking care of a loved one after their hospital stay. Hospital social work moves really fast so the work is usually very concentrated into just a few days, but in that length of time we might meet with a family several times to talk about their options. It is very much a crisis point for the family, and it is important that the family feel as comfortable as possible with the choices they make. Sometimes there are several family members involved, and it is hard for them to come to an agreement about what to do. Some want to offer the family member care at home, and others think it would be best to get a nursing home placement. Meanwhile the clock is ticking toward discharge day. So the social worker can get really stressed about having a plan lined up. The thing that is impressive is that even with all this pressure, the social worker takes her time and tries to talk with the family members together and separately. She engages them as a group in trying to sort out all of the pros and cons and keeps the conversation remarkably positive. She takes every opportunity to comment on how caring the family is toward one another and to point out areas of agreement. She tries to minimize the conflict between the family members by stressing their common concern for the ill family member and other positive things the family has going for them. It's amazing how effective this positive approach can be in neutralizing anger and minimizing conflict.

Working with Groups

Human service professionals often spend much of their day in groups. As an intern, you are likely to see many types of groups that are necessary for the effective operation of the agency as well as groups that are used to meet the needs of clients. Groups that are used within the organization to accomplish administrative goals are generally task groups. Task groups are those that have the primary purpose of accomplishing specific objectives together. This work might include, for example, writing a new personnel policy, planning a community education program, writing a grant to initiate a new community program, or planning a fundraiser. These particular examples of task groups tend to be episodic in nature, and the group generally disbands once the task is accomplished. Many committees and task forces that human service professionals are a part of fall into this category.

Other on-going staff groups are also generally a part of agency life. Examples of such groups include program managers who work together to ensure coordination of various units of the organization, multidisciplinary team meetings that convene to plan and coordinate client care, and supervision groups in which professionals discuss their work and receive feedback from a supervisor and/or one another.

In addition to staff groups, many agencies use groups to meet the needs of their clients. For some organizations, the group is the primary medium for service delivery, whereas for others groups are used as one approach among many. If you are performing your internship in a setting such as a recreation center, community center, afterschool enrichment program, group home, or residential treatment program, you are likely to find that most of your day is spent with client groups. Interns in other organizations may find that client groups are created for diverse purposes. For example, counseling and psychotherapy groups are commonly used in mental health settings, and socialization groups are used to help clients develop behaviors and attitudes that are necessary to be successful in the wider society. As an intern, you might encounter socialization groups in working with delinquent youth or children who have difficulty conforming to the expectations of the school classroom. Recreation groups are routinely used in retirement communities and adult care facilities to help participants build social connections, develop and maintain various skills, and enhance quality of life. Problem-solving or decision-making groups might be used to assist clients in charting a course of action to respond to a problem shared by the group's members. Examples might include residents of low-income housing who are concerned about their living conditions or members of a neighborhood whose streets are becoming increasingly plagued by illegal drugs and other criminal activity.

Group interventions provide a powerful medium for support and change as group members experience the benefits of shared support, brainstorming, and problem solving. Groups can bring energy and motivation to clients who feel overwhelmed and emotionally drained by acting alone. To harness this special power of group intervention, human service professionals must draw upon specialized knowledge and skills. To lead groups effectively, knowledge of group dynamics is needed to guide your work. You must routinely observe and monitor a number of group structures and processes in order to develop an understanding of the group's functioning and their implications for your communication and leadership as a professional. Zastrow (1997) identifies composition, goals, roles, norms, cohesion, leadership, and group development as key elements in any group. These major group concepts are briefly described below along with the significance of each to group functioning.

Composition has to do with the membership of the group. In some situations, the human service professional has no control over group composition whereas in others the professional may have the opportunity to select members who are good candidates for the group's work. For example, clients might be interviewed for a counseling group to ensure that group members are selected who share common concerns and are comfortable with a group approach while also having enough diversity that the group members can benefit from their interactions with one another. Workers in a community recreation facility, in contrast, may work with whatever community members choose to enroll in a given program or show up on a given day. Groups of each type present unique opportunities and challenges.

Goals are necessary in order for groups to have a common focus and direction. Within a group, members may have individual goals as well as group goals. A children's afterschool program with a tutoring emphasis might have a group goal of all students maintaining at least a C+ average collectively in their school grades. Each individual child might also have goals for improving or maintaining her or his own performance, such as improving a grade in reading from C- to C+. A counseling group might have a group goal of all members developing greater assertiveness skills. Some members might focus their individual goals more on family relationships while others' goals might focus more on their work environment.

Roles are normal and predictable parts of group life. Some group members may emerge as task leaders who are effective in getting the work done in the group while others may emerge as maintenance leaders who are effective in developing and maintaining social and emotional bonds in the group. More specific roles such as initiator, compromiser, challenger, troubleshooter, mediator, energizer, tension-reliever, or direction giver emerge as the group develops. There also is always the risk of negative group roles emerging, and the human service professional must try to minimize the risk of this occurring and to curb the damage of such roles if they should develop. Examples of negative group roles include the blocker and distracter. The human services professional must support and employ the role structure of the group to best advantage while also trying to develop and maintain constructive roles within the group.

Norms are the rules and expectations that govern the life of a group. In some groups, the norms seem to be tacitly understood and assumed and become a focus of concern only when one or more members violate the rules that are assumed by the majority of the group. For example, in a staff meeting there may be a tacit norm operating that members should arrive promptly on time. When a few staff members repeatedly arrive late, more explicit discussion of the norms may become necessary in the group. In most groups, especially client groups, it is beneficial to engage the group in setting norms to govern their behavior. In addition to group input, the professional and/or agency may have specific expectations or policies that need to be communicated to the group as well. Norms are necessary to bring a sense of order, safety, and predictability to a group. Research has shown that groups without sufficient norms have a greater risk for group members being hurt or harmed in the group (Schopler & Galinsky, 1981). For this reason alone, it is important that human service professionals exercise care in building adequate group norms.

Cohesion is the degree to which group members feel connected with one another through social and emotional bonds and/or through common concerns and goals. A well-functioning group must have a sufficient level of cohesion to inspire cooperation and mutual caring among the members. Groups with low cohesion may encounter difficulties such as poor attendance, conflict, and competitiveness among the group members. Professionals who are striving to develop groups with higher levels of group cohesion try to minimize competitive activities while emphasizing shared concerns, mutual goals, and common experiences.

Leadership in human service groups is often two-dimensional. On the one hand, there is the professional who initiates the group, convenes the meetings, and provides leadership for the group's formation. Once the group is operating, leadership emerges among the group members as well. Over the lifetime of the group, the professional's leadership might become less instrumental to the group's functioning while the group members assume more leadership. This is a normal and desirable aspect of group development and one that should generally be encouraged by the professional as long as the emerging leadership is positive and appropriate.

Group development occurs in all groups, and many theories have been developed to describe and explain this important group process. One of the more easily remembered and useful theories of group development was put forth by Tuckman (1965) who suggested that groups go through five predictable stages that might be summarized as forming, storming, norming, performing, and adjourning.

1. **Forming:** The group is initiated and the members begin to become acquainted. Members experience some anxiety as they struggle for acceptance and position within the group.

2. **Storming:** Group members resist the control and structure of the group. Members may question whether the group is capable of meeting its goals and whether the benefits of being in the group outweigh its costs. Group members may become competitive with one another or may struggle to redefine the group's purpose and/or methods of operation.

3. **Norming:** Group members reinvest in the group and reshape the group experience to better meet their needs. New rules and expectations are set to govern the work of the group. From this redefinition, group members feel greater ownership of the group and in turn a greater commitment to the group's work.

4. **Performing:** The group is now ready to turn its attention more fully toward goal achievement. Members invest increased levels of energy into the work and develop the skills that are needed to accomplish individual and group goals. Group cohesion is high, and members derive satisfaction from their collective accomplishments.

5. **Adjourning:** The group becomes focused on "moving on" as the work of the group has been accomplished. Members reflect on their shared experience and celebrate their accomplishments. Group cohesion decreases as the work toward common goals no longer provides a group purpose. Group members transition out of the group and/or the group disbands.

In working with groups, the basic communication skills discussed earlier in this chapter continue to be fundamental in your work. Active listening along with the exploration and action skills discussed earlier provide the foundation you need to work with groups effectively. Additionally, human service professionals who work with groups must have the flexibility and skills needed to adapt their roles and communication to the changing needs of the group. Special challenges, such as managing and resolving conflict, handling disruptive behavior, building cohesion, and engaging the group in evaluating its own functioning, are presented by groups.

Corey (2004) discusses specific communication methods and skills for group work, including protecting, blocking, linking, and modeling. *Protecting* involves the professional in intervening to safeguard a member who is being scapegoated, belittled, or treated unfairly. *Blocking* occurs when the professional intervenes to stop counterproductive behavior such as judging, criticizing, excessive advice giving, gossiping, monopolizing the conversation, and conveying misinformation. It is important that blocking is done gently and tactfully and that it addresses the problematic behavior rather than attacking the individual involved. *Linking* is an important skill that helps group members connect with one another. Linking is accomplished by pointing out common themes among the members and encouraging members to talk *to* one another rather than *about* one another. *Modeling* is a method whereby the leader demonstrates the communication skills in the group that the group members need to develop. These skills might include, for example, careful listening, openness, empathy, assertiveness, and use of I-statements. The professional uses I-statements to give feedback to the group in a nonblaming manner. One example of an I-message might be, "I'm getting distracted by the various side conversations that are going on in the group right now. I'd like to ask everyone to focus on Shannon since she has the floor" (Zastrow, 2006).

Group interventions rely on the premise that the forces of group process can be harnessed productively to assist clients and to enhance the functioning of human service agencies and organizations. Your ability to use groups effectively is based on your knowledge of how groups operate and your skills in interacting with groups. You can use the knowledge and skills discussed here to enhance your participation in groups as both a participant and a leader. Your knowledge and skills will be developed more fully as you practice their use in group settings.

Some good organizations exist to support those who are interested in focusing more on group work. These include the Association for Specialists in Group Work and the Association for the Advancement of Social Work with Groups. The websites for these organizations offer extensive resources for enhancing your knowledge and skills in working with groups. The White Stag Leadership Development website also provides extensive information about group leadership and group dynamics.

 ◆ **EXERCISE 6.9 SYNTHESIS: LINKING KNOWLEDGE AND EXPERIENCE**

What groups have you had the opportunity to observe, lead, or participate in during your internship?

As you think about these groups, what examples do you see of group structure and development as discussed in this chapter?

When have you (or others) used the communication skills described above?

What do you see as your strengths in working with groups?

What are your goals for future improvement?

A Student's Reflections on Working with Groups

The young mom's group is always one of the highlights of my week. The Family Center has a group of teen moms that meets every Tuesday afternoon. There are eight members of the group between 14 and 18 years of age. All of them have at least one child 2 years old or younger. There is always a short educational part of the group time followed by open discussion. The topics have ranged from juggling school with being a parent, parenting skills, budgeting, information about child development, how to maintain a social life, etc. The discussion time is the most interesting part of the group. These young women have a lot in common and seem to really enjoy being together. My supervisor does a great job of drawing everyone into the conversation and also makes sure no one monopolizes the group time. Childcare is provided during the group, and the moms always talk about how great it is to have an uninterrupted conversation. Each of the members has set some personal goals to work on, and the

continued

group's overall goal is to support one another in their efforts to finish high school and be a good parent. I think the most challenging part of leading this group is that each of the members has a lot of stress to deal with and is pretty needy. It is hard to give everyone the attention they want and need sometimes. It is clear that the group is badly needed. It would be great if they could get this support every day.

Working with Communities

In addition to working with individuals, families, and groups, human service professionals also intervene with communities. The community may be viewed as a local neighborhood, a town, city, state, or even the broader society of a country or the global community. In working with communities, the focus is on changing the environment in which people live in order to improve their quality of life. As a thoughtful human service intern, you will probably recognize conditions that need to be corrected or improved in order for your clients to experience significant improvement in their lives. For example, working in a homeless shelter will surely raise questions about what conditions in the community are creating homelessness. Has there been an economic downturn that prompted layoffs and cut-backs? Have companies pulled out of the area to relocate in countries with less expensive labor? Have other human service organizations cut back services that were needed to help people maintain their ability to work and pay rent? Has the local population outgrown the availability of low-income housing?

Working with communities requires that you take a step back from providing direct services day to day to ask yourself what conditions create the need for these services and how these conditions might be addressed. By making such thinking a part of your reflective practice throughout your career, you can play a role in addressing systemic problems that contribute toward individual problems and needs. Some human service professionals choose to become fully involved in community practice, making it the central focus of their careers. Here we examine three fairly common areas of community practice: community education, community organization, and political engagement.

Community Education

Community education engages professionals in educating citizens about issues relevant to their lives. As a community educator, you may make presentations before community groups to raise awareness on a given issue and to inform participants how they can become involved. To perform community education effectively, you need to be able to assess the needs and interests of the community or audience, research a topic, prepare educational materials such as handouts and visual aids, and speak before a group with confidence and comfort. Also, your ability to engage an audience in the topic is key to your effectiveness as a community educator.

Many community agencies prepare presentations regarding the issues they deal with and are eager to share this information with local groups. By the same token, community groups often contact agencies with requests for presentations on a given topic. For example, an organization that serves older adults may make community presentations about Alzheimer's disease and other forms of dementia in an effort to help community members make lifestyle choices that are associated with healthy aging and help them recognize these conditions early so that medical intervention can be sought. Similarly, an organization that assists children with developmental disabilities might provide education about the local services available to parents of children with disabilities while also providing prevention education to prospective parents about prenatal practices that increase the likelihood of healthy births.

Much of community education is directed toward problem prevention. Chapter 9, Writing and Reporting Within Your Field Agency, discusses the skills involved in making professional presentations. This information will be helpful as you develop your skills in community education.

Community Organization

Community organization involves mobilizing citizens to address issues that affect their lives. Such organizations can occur on the local level when neighborhoods address problems such as cleaning up streets or waterways, blocking an unwanted road that threatens to divide a community, or organizing a community center to provide a safe place for children to gather after school. Community organization can also occur on a much larger scale. Citizens of a city can mobilize to develop a vision for their future, to create a plan for economic development, or to initiate a new human service program to address an unmet need. Citizens of a region might mobilize to investigate and locate the sources of the acid rain that affects their living environment. Citizens of a state might mobilize to expand affordable health care services in rural areas throughout the state or to improve mental health care.

Community organization can even occur on the national level as like-minded citizens band together to accomplish a common goal such as securing prescription drug coverage for older adults. Broad social movements such as civil rights and women's suffrage were accomplished in part through community organization of this type as it occurred across the country (Brueggemann, 2006).

The communication skills involved in mobilizing communities draw upon many of the skills discussed previously in this chapter, especially those in group dynamics and leadership. Numerous additional skills will come into play as well—public speaking, persuasive writing, forming coalitions between groups with similar interests, and communicating through websites, e-mail, and other methods of reaching large groups (Homan, 2004).

Political Engagement

As human service professionals understand the relationship between community needs and individual problems, they tend to become more involved in trying to address those community problems. At times, both professionals and the citizens they work with recognize that there are limits to what they can accomplish on their own. Sometimes legislation is needed to create change. Political engagement then becomes a necessary step in creating community change.

At minimum, political engagement involves voting for candidates who understand the issues that you are concerned about and who will probably support the changes that you are working toward. Political engagement also involves working with elected representatives to advocate for the needed changes and includes activities such as lobbying for increased resources to fund services that the community needs or lobbying for legislation to protect the rights of citizens in various situations. At times, such legislation might already exist, but the current laws are not being enforced. In these situations, using the courts to enforce these laws can be an appropriate and useful approach for community workers to use in promoting change (Homan, 2004).

Communicating effectively to perform advocacy through political engagement requires a number of skills. In advocating for your issue, you might speak with individual legislators or community members or you might address large groups of people, influencing them to rally around the cause. Your ability to connect with people one to one as well as in groups will be important skills for this level of community work. You must be able to clearly articulate the issue that you want to address and to discuss it exhaustively. Your knowledge of the history and background of the issue must be wide ranging, based not only on personal anecdotes but also on extensive research.

Gathering facts and statistics is an important, indispensable part of this process, but it is equally important that you are able to give your issue a human face, conveying in a short, compelling story how a specific person was affected by the issue. The story should be crafted in such a way that the listener feels empathy for the person and is motivated to action. Selecting effective spokespersons for the cause is also important. It is, of course, most effective to have spokespersons who have been personally and directly affected by the issue (Community and Government Relations, 2005).

The skills required for successful community intervention are many and varied. You might understandably feel anxious about tackling problems of a larger scope if you are more accustomed to working with individuals, families, and small groups. Homan (2004) points out that "the fundamental skills you need to work at the individual level are the same skills you use when you work for community change" (p. 82). Your ability to form relationships, assess needs, analyze systems, set goals, plan and evaluate interventions, facilitate and lead groups, work with diverse populations, and link people with common concerns all come into play when the focus of change is larger systems, just as they do when the focus of change is the individual or smaller systems.

There are numerous websites devoted to working with communities that can be useful to you as you encounter various community needs. The Community Toolbox, a website provided by the University of Kansas, offers extensive resources for community work of different types. Also, the Voluntary Sector Knowledge Network website provides information about nonprofit organizations, including community and government relations.

Many professional organizations for community workers, such as the Association for Community Organizing and Social Administration, the National Organizers Alliance, and the Community Development Society, also provide websites with information ranging from current policy initiatives in need of support to skill building resources. Although most human service professionals do not lead national or state movements, there are always many opportunities to support and participate in such efforts for the well-being of the people you serve, and even for your own well-being. Most local communities also have many issues that you can become involved in or even lead as a human service professional and citizen. Participating in such initiatives can be a gratifying way to improve the quality of life in your community. The most important step is simply choosing to get involved. Your skills will improve over time as you practice the art and science of community change.

A Student's Reflections

My agency recently received a grant to carry out programs to prevent childhood sexual abuse in the county. They were successful in getting the grant because our county has one of the highest rates of childhood sexual abuse in the state. They are approaching their task with a lot of community input and involvement. They are meeting with different focus groups around the county to get input and ideas about effective ways to prevent childhood sexual abuse. They have met with groups of pastors and religious leaders, school personnel, medical professionals, human service professionals, etc. They are also meeting with client groups in different agencies. Right now they are even trying to get into some community groups like civic groups, books clubs, church groups, and such for input. A lot of good ideas are surfacing, and they are asking each of these groups what they could do to help with the effort. It is building a lot of community buy-in. I had always thought of community education as just going around making presentations to groups or doing health fairs or whatever. I see now how much more effective it is to work with a community in a deeper way in order to get real results.

 ◆ **EXERCISE 6.10 SYNTHESIS: LINKING KNOWLEDGE AND EXPERIENCE**

What community problems affect the populations you work with in your internship?

In what ways, if any, is your agency involved in addressing community problems through community education, community organizing, or political engagement?

What further community actions do you see that your agency might take to improve the quality of life for its clients?

Conclusion

Communication is the most important skill that you will employ in your human service internship and career. Communicating with individuals, families, groups, and communities will most likely be the primary focus of the typical day in your field site. Reviewing your knowledge about communication and applying this knowledge to various practice situations will enhance your skills dramatically during the course of your internship. It is important to reflect on the quality of your communication continuously, setting goals for further improvement.

◆ FOR YOUR E-PORTFOLIO

Discuss your experiences thus far in working with individuals, families, groups, and communities. Which of these levels of intervention have you had the most experience with? Which have you had the least experience with? If you were employed in your internship organization, how might you expand your work to encompass all four levels of intervention? Assess your abilities in each of these levels of intervention, identifying your strengths as well as areas that need improvement. As you enter the human service profession, which area(s) of intervention is (are) most interesting or satisfying to you and why?

References

Brill, N. (1995). *Working with people: The helping process*. White Plains, NY: Longman.

Brueggemann, W. (2006). *The practice of macro social work* (3rd ed.). Belmont, CA: Thomson.

Capuzzi, D., & Gross, D. (2007). *Counseling and psychotherapy: Theories and interventions* (5th ed.). Engelwood Cliffs, NJ: Prentice Hall.

Community and Government Relations. (2005). Voluntary Sector Knowledge Network. Retrieved December 20, 2005, from http://www.vskn.ca/commune.htm.

Corey, G. (2004). *The theory and practice of group counseling.* Belmont, CA: Brooks/Cole.

Egan, G. (2002). *The skilled helper: A problem-management and opportunity development approach to helping.* Pacific Grove, CA: Brooks/Cole.

Egan, G. (2006). *Essentials of skilled helping: Managing problems, developing opportunities.* Belmont, CA: Thomson.

Ezell, M. (2001). *Advocacy in the human services.* Belmont, CA: Wadsworth.

Goldenberg, I., & Goldenberg, H. (2000). *Family therapy: An overview* (5th ed.). Belmont, CA: Wadsworth.

Hanna, S. & Brown, J. (2004). *The practice of family therapy: Key elements across models.* Belmont, CA: Thomson Brooks/Cole.

Homan, M. (2004). *Promoting community change: Making it happen in the real world* (3rd ed.). Belmont, CA: Wadsworth/Thomson.

Mandell, B., & Schram, B. (2006). *An introduction to human services: Policy and practice.* Boston: Allyn and Bacon.

Meier, S., & Davis, S. (2005). *The elements of counseling* (5th ed.). Belmont, CA: Thomson Brooks/Cole.

Minuchin, S. (1974). *Families and family therapy.* Cambridge, MA: Harvard University.

Neukrug, E. (2002). *Skills and techniques for human service professionals: Counseling environment, helping skills, treatment issues.* Pacific Grove, CA: Brooks/Cole.

Neukrug, E. (2004). *Theory, practice, and trends in human services* (3rd ed.). Pacific Grove, CA: Brooks/Cole.

Okun, B. (2002). *Effective helping: Interviewing and counseling techniques.* Pacific Grove, CA: Brooks/Cole.

Patterson, L., & Welfel, E. (2000). *The counseling process* (5th ed.). Stamford, CT: Wadsworth.

Petr, C. (1998). *Social work with children and their families: Pragmatic foundations.* NY: Oxford University Press.

Ragg, D. (2006). *Building family practice skills: Methods, strategies, and tools.* Belmont, CA: Thomson.

Rooney, R. (1992). *Strategies for work with involuntary clients.* New York: Columbia University Press.

Rose, S. (1992). *Case management & social work practice.* White Plains, NY: Longman.

Schopler, J., & Galinsky, M. (1981). When groups go wrong. *Social work, 26,* 424–429.

Thomlison, B. (2002). *Family assessment handbook: An introductory practice guide to family assessment and intervention.* Pacific Grove, CA: Brooks/Cole.

Tuckman, B. (1965). Developmental sequence in small groups. *Psychological Bulletin, 63,* 384–399.

Vourlekis, B., & Greene, R. (1992). *Social work case management.* Hawthorne, NY: Aldine De Gruyter.

Welfel, E., & Patterson, L. (2005). *The counseling process: A multitheoretical integrative approach* (6th ed.). Belmont, CA: Thomson Brooks/Cole.

Woodside, M., & McClam, T. (2006). *Generalist case management: A method of human service delivery* (3rd ed.). Belmont, CA: Thomson Brooks/Cole.

Young, M., & Long, L. (1998). *Counseling and therapy for couples.* Belmont, CA: Wadsworth.

Zastrow, C. (1997). *Social work with groups* (4th ed.). Chicago: Nelson-Hall.

Zastrow, C. (2006). *Social work with groups: A comprehensive workbook* (6th ed.). Belmont, CA: Thomson Brooks/Cole.

Chapter 7

Dealing with Diversity

Internships generally give students the opportunity to interact with people very different from themselves, people representing a range of cultural groups. This chapter guides you through several important principles and skills for working with diverse groups. The exercises encourage you to apply these principles and skills to your fieldwork experiences.

A Student's Reflections on Diversity

One of my biggest surprises about this school is how diverse the student body is. First of all, there is socioeconomic diversity. The school draws from parts of town with the wealthiest and the poorest children in the county. Also, the children come from all over the place. They range from White to African American to Hispanic to Laotian to Vietnamese. There are even some European children here—two students who recently moved here from Sweden and several children from five different German families who recently moved here after a local company was bought out by a German company.

An Overview

According to the U.S. Census Bureau, the population of the United States is becoming ever more diverse. Based on the 2000 census, projections are that Black, Hispanic, Asian, and Native American segments of the population will grow significantly over the next 50 years. Non-Hispanic Whites are predicted to comprise a diminishing percentage of the total population over this period. Between 2006 and 2010, Asian and Latino populations are projected to increase most dramatically (Robinson, 2005). These demographic shifts are occurring due to differential birthrates among the various groups and to immigration patterns.

Those 65 years of age and older are one of the fastest growing groups in the country. The median age of Americans is expected to increase through 2030 due to the growing population of elderly. Adding to the diversity picture, the number of people with disabilities is also increasing (Day, 2000). There is also great religious diversity in the United States. Although Christianity is the most prevalent religion, human service professionals are also likely to work with clients who are Jewish, Buddhist, Hindu, or Muslim as well as those who are not religious or who define their spirituality outside the context of any traditional religion or faith (Robinson, 2005).

One of the greatest challenges facing human service professionals being educated today is working effectively with individuals, families, and groups from highly diverse backgrounds and cultures (Neukrug, 2004). Due to the growing diversity of the U.S. population, the relationships between human service professionals and their clients are becoming increasingly cross-cultural, that is, the client is from one cultural background and the worker from another. Such relationships present special challenges for the human services professional because the worker and client are likely to bring very different assumptions, values, beliefs, and communication patterns to the helping relationship.

Sue, Ivey, and Pederson (1996) assert that the educational preparation of human service professionals must be radically altered to bring cultural consciousness to a more central focus in this training. According to their model, cultural considerations should be a part of a professional's mindset in working with every client, not just an "add-on" that is applied when working with those who are obviously from a "different culture." This chapter's approach is consistent with this idea because it encourages you to think about cultural issues more broadly, considering not only ethnicity but also age, gender, sexual orientation, rural/urban/suburban locale, socioeconomic class, and regionalism (North, South, Western, Midwestern, New England, etc.) among others.

Understanding Concepts Related to Diversity

For individuals within a given environment or group, acquiring the characteristics of that culture is generally adaptive, helping them to cope and meet their needs within that particular social setting. Culture has been defined as the "knowledge, language, values, customs, and material objects that are passed from person to person and from generation to generation in a human group or society" (Kendall, 2006, p. 43). Cultural characteristics include such variables as language and communication patterns (both verbal and nonverbal), attitudes, norms, values, relationship and kinship patterns, and religious beliefs (Cox & Ephross, 1998). Culture can be a useful concept in understanding not only ethnic differences but also the more commonly experienced differences associated with gender (Bricker-Jenkins & Lockett, 1995) and socioeconomic class (Schlesinger & Devore, 1995).

Closely related to the concept of culture is that of ethnicity. Ethnic groups may be defined as a particular type of cultural group—that is, "a collection of people distinguished, by others or by themselves, primarily on the basis of cultural or nationality characteristics" (Kendall, 2006, p. 544). People within a given ethnic group generally perceive themselves, or are perceived by others, as sharing a common history and/or ancestry. Some writers also point out that the term is generally used to denote a group of people who maintain a shared culture within an environment in which a different culture is dominant (De Vos, 1975). Thus, Anglo-Americans, even though constituting a cultural group, would not be seen as an ethnic group in the United States due to their dominance within the larger culture.

Another important concept to keep in mind as you try to understand and work within a culturally diverse population is that of minority groups. A minority group, sometimes referred to as a subordinate group, may be defined as "a group whose members, because of physical or cultural characteristics, are disadvantaged and subjected to unequal treatment by the dominant group and who regard themselves as objects of collective discrimination" (Kendall, 2006, p. 282). This definition implies that minority groups stand in contrast to "dominant groups." A dominant group is one that "is advantaged and has superior resources and rights in society" (Kendall, 2006, p. 281). As these definitions suggest, minority group status is based on the relative powerlessness of a group in relation to the dominant group(s) within the larger society. Minority status therefore is not a reflection of a group's size but of its relative power. Therefore, women, because they hold collectively less power and fewer resources, might be considered a minority group, even though they comprise roughly half the population.

Although it is important to understand these important concepts related to culture, preparing yourself to work effectively with diverse populations, of course, requires far more than this. To become capable of working with diverse groups effectively, people must engage in the intentional building of competence and skills as cross-cultural helpers. Some basic cross-cultural competencies include gaining knowledge of other cultures through study and direct experience, developing a greater awareness of your own cultural make-up, decreasing your ethnocentrism, and increasing your respect of other traditions (Lum, 1999; Petrie, 1999). Furthermore, some writers suggest that effective cross-cultural helpers can benefit from understanding biculturalism as a potential coping skill for clients who are struggling to adapt within cultures different from their own (Chau, 1991; de Anda, 1984; Lukes & Land, 1990; Schram & Mandell, 1994; Valentine, 1971). Finally, and most important, the human service professional must be able to apply such knowledge effectively and skillfully in actual practice situations (Weaver, 1998). This chapter helps you to begin developing some of these skills as you think about issues of diversity related to your fieldwork experience.

Experiencing Diverse Groups

◆ EXERCISE 7.1 PERSONAL REFLECTION: OBSERVATION OF SELF AND OTHERS

You have probably had the opportunity to observe and work with individuals from diverse groups during your field experience. This exercise will help you develop a clearer picture of this diversity. Using the prompts below as your starting point, describe the populations with which you are working during your field experience.

"Through my fieldwork I have the opportunity to observe or interact with individuals representing the following . . .

age group(s):

ethnic group(s):

gender group(s):

minority group(s):

racial group(s)

region(s) of the country:

religion/spirituality and/or belief systems:

sexual orientation(s):

socioeconomic group(s):

Review and reflect upon what you have just written.

How do the people being served within your agency differ from one another?

How do the people being served in your organization differ from you?

Increasing Your Knowledge About Other Cultures

As you work with people from varied backgrounds, it is sometimes difficult to sort out those characteristics that are cultural traits as compared to those that are individual traits. Within all cultures, there is a wide range of individual diversity. Groups that might superficially seem to be homogeneous are often surprisingly diverse. Hispanic or Latino people, for example, may appear to have similar backgrounds, but this is in fact a very diverse group. A Latino might have cultural roots in Cuba, Mexico, South America, Central America, Spain, or Puerto Rico. Each of these groups constitutes a distinct subgroup within the broader Latino ethnic group, and each has its own history and cultural identity (Longres, 1995).

To assume that all people within a given cultural group are alike or even similar is to stereotype. Although stereotyping is to be avoided, general information about a client's background and cultural identity is often useful. You cannot and should not assume that your clients conform to common characteristics within their culture, but having a general knowledge of their culture will help prepare you to understand perspectives and world views that might be different from your own. Having a better understanding of a client's culture enables you to work with them in a more sensitive manner and possibly helps you to understand the origins of certain behaviors that you might otherwise misinterpret. For these reasons, human service professionals have a responsibility to learn about the cultures that their clients represent (Lum, 1999).

Using the information generated in Exercise 7.1, identify the various groups or cultures that you need to learn more about in order to work knowledgeably and sensitively with your agency's clients. If the goal of learning more about these groups is not already part of your learning plan, consider adding this goal, along with strategies to achieve it. For example, reading about or interviewing someone from these cultures will enable you to learn more about the culture.

Out of your increased understanding of the culture, you will be better prepared to recognize culturally learned behaviors as well as individual differences among your clients. Most important, you will be better equipped to develop helping relationships with your clients built upon acceptance and understanding of their cultural context. Clients themselves are indispensable sources of information about their culture. Listening to clients with an attitude of openness and respect is perhaps the most important thing you can do to overcome cultural barriers in your helping relationships (Halley, Kopp, & Austin, 1998). The acceptance and respect that you convey through your interest in their unique backgrounds contribute immeasurably toward developing a positive and productive cross-cultural helping relationship.

As you read about various cultures and interact with clients from different cultural backgrounds, look for the strengths and assets within each culture (Lum, 1999). In many human services settings, it is easy to become focused exclusively on needs and deficiencies. Such an orientation may cause you to overlook the resourcefulness, resiliency, strength, and capabilities that are part of each culture. Recognizing cultural assets will help you to avoid negative stereotyping, especially toward groups typically thought of as "disadvantaged" (Nerney, 1998).

In addition to focusing on strengths, it is useful as well to focus on the similarities and commonalities between yourself and those clients who, on the surface, seem to be quite different from you (Nerney, 1998). Although cultural competence requires you to focus on cultural differences, it is essential not to lose sight of those universal human qualities that transcend culture as well. Developing cultural competence means that you develop the ability to recognize and empathize with universal human feelings, needs, and desires, despite the cultural barriers and challenges that various relationships might present. Understanding how these universal qualities tend to be expressed within different cultures will help you to recognize these qualities more readily in cross-cultural relationships. Recognizing the commonalities between you and the clients you serve will enable you to relate more genuinely, empathically, and effectively with clients from all backgrounds.

◆ EXERCISE 7.2 PERSONAL REFLECTION: OBSERVATION OF SELF AND OTHERS

Think about the various cultures that you have had an opportunity to work with during your internship. (You might find it helpful to review your work in Exercise 7.1 for an overview of these group.) What particular strengths or assets can you identify within these various cultural groups?

Think about any particular individuals whom you have encountered during your fieldwork who seem to be very different from you. Also, think about cultures that you have encountered that seem to be very different from your own cultural group. Try to look past the more obvious differences to see the commonalties that are present. What commonalities are you able to observe between yourself and the other person? Between your culture and other cultures you have experienced during your fieldwork?

Developing Awareness of Your Own Culture

All of us see the world through the lens of our own unique culture, although most of the time we are not aware of this lens. Becoming aware of the cultural lens through which you view the world is perhaps one of the most significant, and most difficult, aspects of working effectively with those from other cultures (Lum, 1999). In developing competence as a cross-cultural helper, exposure to people from other cultures is necessary, not only because it enables you to learn about other cultures but also because it enables you to learn about your own culture and become more aware of your own cultural make-up.

In fact, awareness of your own cultural make-up is often best developed through directly experiencing contrasting cultures (Axelson, 1993; Corey, Corey, & Callanan, 2003; Lee, 1991). One of the benefits of interacting with people of different cultural backgrounds is that it heightens your awareness of differences in attitudes and behaviors. Much of your cultural learning occurred so early in your life that it probably exists outside your awareness. Therefore, unless you experience cultural contrasts, you are likely to take your own culture for granted, seeing it as what is "right" or "natural," particularly if you are from the dominant culture. When your culture is all around you, it tends to be like the air you breathe—all-important but outside your awareness.

The importance of gaining cultural awareness through interacting with contrasting cultures is not limited to the dominant cultural group. According to Axelson, "a

Black person in the presence of a group composed mostly of White members might be more aware of his or her Black ethnicity than in a group composed mostly of Black members" (1993, p. 32). Whatever your cultural heritage and identity, experiencing cultural contrast is an important element in developing greater awareness of your own culture. The diversity that you experience during your fieldwork probably offers multiple opportunities to experience this contrast.

◆ EXERCISE 7.3 SYNTHESIS: LINKING KNOWLEDGE AND EXPERIENCE

Below are listed some commonly cited cultural variables—that is, some important components of culture that vary from one group to another. Although you might not have observed all of these variables in your internship, you probably have observed some of them. Which of these cultural variables have you been able to observe in your internship? Describe what you have observed about your own cultural make-up in contrast to other cultures you have experienced in your fieldwork in relation to these variables.

Sense of self, including assumptions about autonomy and responsibility to others

Communication and language, including verbal and nonverbal expression

Understanding of and relationship to authority

Spirituality/Religious beliefs and/or practices

Use of space

Use of time

Food and customs around eating

Attitudes toward work

Definitions of success

Beliefs about health and healing

Beliefs about giving and receiving help

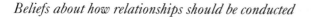

Beliefs about how relationships should be conducted

The value placed on appearance and hygiene

Beliefs about education and the value of education

A Student's Reflections on Diversity

In my family we rely on traditional Chinese medicine. Growing up I saw American trained doctors at times but the main way we thought about health in my family was from the Chinese culture. Chinese medicine looks at the whole person and not just your symptoms. In Chinese medicine spirituality, emotions, diet, herbs, exercise and other things are all considered as ways to help people stay healthy or get better. It is the doctor's job to keep you well, not just to help you when you are sick. Chinese medicine looks at all of what is going on with a person and does not just treat the symptoms. All through my human service education, I have learned that the profession tries to look at the whole person. But when I am out in the field, I often feel that the service providers are not always looking at the whole person. Sometimes the focus becomes very intense on just solving the problem and I find myself thinking, "This isn't going to work." I seem to be the one that's out of step because the American way is to focus on the problem and solve the problem, and I am always wanting to take a broader look. I think this is maybe because of my Chinese background.

Decreasing Ethnocentrism

The first steps in developing cross-cultural competence are learning about other cultures and becoming aware of your own cultural background, biases, stereotypes, and values (Neukrug, 2004). Through these initial steps, you have developed an understanding that cultural traits are simply learned behaviors. Nevertheless, there is a tendency to see one's own culture as "right," "normal," or "natural" and other cultures as "wrong," "abnormal," or "unnatural." This tendency is a manifestation of a phenomenon known as *ethnocentrism*. Ethnocentrism occurs when one assumes that one's own culture sets the standard against which other cultures should be evaluated, that one's own cultural patterns are best and right (Kendall, 2006).

In order to work effectively with people from different cultures, human service professionals must strive to overcome ethnocentrism. By examining their assumptions and reactions as they interact with people very different from themselves, professionals can become more aware of their ethnocentric reactions. Most, if not all, human service professionals have had thoughts and feelings in the course of their work that have brought them face to face with their own ethnocentrism. Ethnocentric reactions might be experienced as feelings of rejection, judgmental thoughts or feelings, or a sense of shock or even disgust in reaction to a client's behavior, attitudes, or lifestyle. These can be difficult experiences, especially for people who generally see themselves as accepting and nonjudgmental.

It can be difficult to recognize ethnocentrism in ourselves, just as it is difficult at times to recognize our own cultural make-up. Each person's cultural make-up might be thought of as a lens through which he or she views the world. This lens is an important element in ethnocentric reactions as it acts as a powerful filter on our perceptions, causing each of us to pick up on certain things in our environment while remaining oblivious to others. This lens also colors how we interpret what we observe, assigning meanings that may or may not be accurate or appropriate to the reality of the situation. The following example illustrates how different cultural lenses operated to give two students very different perceptions of and reactions to the same field setting. Each of their reactions might be considered an ethnocentric response.

CASE EXAMPLE: ONE PLACEMENT + TWO STUDENTS = CULTURAL INSIGHT

Gamal and Sherry were human services students placed in the same child development center for their fieldwork. Gamal was from a working-class family that originated in the Middle East. Sherry was reared in an upper-middle-class family from the northeastern United States. Although the two students were in the same setting at the same time, they had vastly different reactions to and concerns about the children and families served by the child development center.

Gamal had requested a placement in a child development center, despite his family's strong views about the importance of traditional gender roles. In spite of Gamal's flexibility in thinking about his own roles, he found that he was uncomfortable when he saw little boys in the classroom play with dolls or with the play kitchen. In fact, he was so uncomfortable that he sometimes approached these children to initiate a different form of play. Gamal was also amazed by how much "stuff" all the children seemed to have—toys, books, games, records, videos. Not only did the center possess these items in vast supply, but the children brought even more of them from home. Gamal sometimes felt overwhelmed by all of it and frankly felt that it was very wasteful and extravagant.

Sherry, who was in this same setting, had not thought twice about the little boys playing with the dolls and the kitchen, nor had she really thought about the supplies, materials, and equipment within the center. All of this seemed quite natural to her. She was, however, very concerned about some of the children. She was upset to find that some children were left at the center for 10 to 11 hours per day, feeling that this was far too long for a child to be in a day-care center each day. She was also concerned about seeing some children wearing the same clothes to the center two or three days in a row. She saw this as evidence that the children were not being properly cared for.

Gamal and Sherry worked under the supervision of the same field supervisor and talked with her independently about their observations and feelings. The supervisor noted their markedly different reactions to their experiences and recognized the real learning opportunity that was present in getting the two interns together for a discussion. They were amazed at one another's reactions to the center, and each was able to give the other a different perspective on the issues that concerned them. Most interesting of all, neither student recognized that he or she was having an ethnocentric response to the situation until they met together. Only through this discussion did the two students recognize how their own cultural make-up—that is, their own cultural lenses—colored their perceptions of their experiences in their fieldwork.

◆ **EXERCISE 7.4 SYNTHESIS: LINKING KNOWLEDGE AND EXPERIENCE**

Think carefully about the discussion of ethnocentrism as well as the case example. In what ways might your cultural lens have colored your reactions to your field agency and/or the clients it serves?

Describe at least one experience or incident in your fieldwork in which you experienced an ethnocentric reaction. What thoughts and feelings did you have in that situation? As you describe the incident, explain the specific aspects of your own cultural experience that played into your reaction.

A high degree of self-awareness, maturity, and openness to personal growth is required to reduce ethnocentrism and become more accepting of other ways of being. In fact, several aspects of self-actualization identified by Maslow (1954) reflect the types of developmental shifts that are integral to overcoming ethnocentrism. Among other traits, Maslow describes the ideal, healthy, fully functioning person as one who is more accepting of self and others, more perceptive of reality, more independent of culture and environment, more resistant to cultural influences, and more capable of transcending cultural bias. All of these traits, according to Maslow, are central to achieving the highest levels of personal growth and development. Interestingly, these same traits will help you become more effective as a human service professional as well, enabling you to work more sensitively with culturally diverse groups.

As you strive to overcome ethnocentrism and develop respect for different cultures, it is helpful to remember the adaptive nature of culture for people in a given environment. It will be helpful to ask yourself, "How might this particular characteristic have helped this individual to thrive, cope, or survive within the environment in which he or she has lived?" Especially challenging for human service professionals is the fact that at times their clients are hampered by their culturally learned behaviors. Although certain behaviors are adaptive and desirable within their own environments, the client may experience distress in certain situations because these same behaviors violate the norms of the dominant culture, creating conflict or stress. The client then is in the confusing situation in which a behavior that is constructive and positive in one setting is counterproductive in another setting.

For example, the dominant Anglo culture of the United States places a high value on punctuality. One manifestation of this cultural value is that businesses and other organizations generally operate on a system of precise schedules. Some cultures, for example, Appalachian mountain culture, manage time far more loosely (Axelson, 1999). Individuals from such cultures may experience severe disapproval if they bring these habits regarding time into the typical American workplace. Such situations, often referred to as *culture clashes*, can seriously affect an individual's ability to experience satisfaction and success within a different culture. An individual's ability to hold a job and meet his or her own survival needs might be significantly hampered in such a situation.

Understanding Biculturalism as an Adaptive Mechanism

As our country becomes more culturally diverse, human service professionals must often help clients cope with culturally challenging situations in their own lives. The worker's role in these situations is, as in all helping encounters, to help the client explore and understand the nature of their difficulty and identify ways that it might be resolved. Some clients in such situations can feel quite threatened because they fear that they are being called upon to give up their own cultural identity in order to be successful. Other clients may discount the value of their own culture and see it only as an obstacle to be overcome if they are to be successful. The notion of biculturalism is sometimes helpful to people in such situations.

Biculturalism occurs when an individual is able to adapt his own behavior to a particular culture as needed while retaining his primary cultural identity. The fully bicultural person is able to shift successfully back and forth between cultures, experiencing gratification of needs within, as well as making contributions to, both worlds (Chau, 1991; de Anda, 1984; Lukes & Land, 1990; Valentine, 1971; Van Den Bergh, 1991).

The major benefit of biculturalism as a coping strategy is that it respects the client's allegiance to and identification with her primary culture, recognizing the value of the primary culture to the client's identity and ability to cope within that social environment. At the same time, biculturalism allows room for the client to adapt to another culture as necessary in order to meet her needs within that environment as well. Within the framework of biculturalism, clients can experience success in new or different cultures while maintaining their primary cultural identity. Although this adaptation process might require a good bit of effort and energy, it tends to be less stressful than the alternatives of giving up the primary culture altogether or experiencing repeated frustrations and difficulties within the different cultural setting. The following case example illustrates such a situation.

CASE EXAMPLE: A CLIENT LEARNS TO BE BICULTURAL

Maria, a 33-year-old, single, Italian-American woman, grew up in an ethnic community in which her family operated a family business for many years. The norm within that community and within her family was an "all-for-one-and-one-for-all" mentality. Individual needs and wants were seen as not being as important as the needs of the group. There was a high value placed on hard work and each and every person carrying his or her own load without complaining. If a person was asked to help another, then compliance was automatic. Within this context, it was assumed that people were always available to help one another and that if people asked for help, they really needed it. Maria attended a local university while continuing to live at home and work in the family business. After graduation, she accepted a position in a large corporation and continued to live at home, working in the family business on weekends. Maria took her cultural learning about working within a group to her new position. She was bright, capable, hard-working, and a team player, all of which placed her in high regard within this corporate setting.

Despite its obvious strengths, Maria's cultural learning had left her with some significant gaps in her ability to cope in the corporate world. She had not learned to be assertive and set reasonable limits regarding her workload. In fact, within her culture, to do so would have been seen as selfish and irresponsible. Maria also had not learned how to take credit for her work. She repeatedly was in the position of doing all or part of another worker's project because he or she asked for help but then received no credit for this work as her co-worker in the end reaped the rewards. Maria did not feel that she could say "no" to her co-workers' requests for her assistance and did not have the assertiveness to even address what was going on with her co-workers or supervisor. She sought help when she had become very frustrated and angry about her situation.

A culturally attuned helper was able to help Maria see her situation in its cross-cultural context. Behaviors that worked for her in her ethnic community, based upon values that she prized, did not work as well in the corporate setting. Maria made some decisions about whether she wanted to stay in the corporate setting and evaluated what she would need to do to meet her own needs within that setting. Her human services worker helped her to see that she could choose, if she wished, to work toward acquiring behaviors more adaptive to the corporate culture (e.g., limit-setting, assertiveness, and calling attention appropriately to her personal accomplishments), while maintaining certain behaviors of her primary culture. While recognizing the adaptations she might need to make in order to find more satisfaction in her job, Maria also gained a clearer appreciation of the strengths of her primary culture, recognizing how its work ethic and emphasis on teamwork were helping her to meet the challenges of the corporate world.

Although it took some time and energy, Maria learned to be bicultural. Over time, she became able to shift between the corporate world and her ethnic community with relative ease. While she remained most comfortable within her primary culture, she felt that her adaptation to the corporate culture had given her an opportunity to grow in ways that she never would have done otherwise.

Like Maria, many clients from various ethnic groups, lower socioeconomic groups, or non-Western cultures often feel strongly bound to the consideration of group interests and less bound to self-interest in their decision making (Ewalt & Mokuau, 1995; Petrie, 1999). This phenomenon has also been noted as a common one among women as compared to men (Gilligan, 1982). This issue is but one of many cultural patterns that can create challenges for individuals as they function in the dominant culture while maintaining their own cultural identity.

◆ EXERCISE 7.5 SYNTHESIS: LINKING KNOWLEDGE AND EXPERIENCE

Think about the clients with whom you have worked during your internship. Describe how one or more of these clients might be struggling with conflicting expectations and standards as they move between two cultures. How might the idea of biculturalism be useful in helping these individuals cope more successfully?

Not all clients will be open to the idea of bicultural adaptation, and biculturalism certainly cannot be seen as a panacea. For those clients looking for ways to adapt within culturally different environments, biculturalism can be a useful approach.

Understanding Social Location

Human service education routinely includes the study of various cultures that professionals encounter working in American communities. Although this approach is considered fundamental to developing cultural competence, it has also been a source of concern to some human service professionals due to its "assumption of whiteness" (Yee, 2005). That is, cultural diversity training models that acquaint students with various cultures introduce these cultures as the "other" or the "unfamiliar." "White" culture is assumed to be dominant and is assumed to be the culture of the service provider. In reality, human service professionals come from a variety of cultural backgrounds.

Although much attention is given to the White professional working with minority clients, less attention is generally given to the minority professional who works with clients from the dominant, or White, social group. Issues of power, racism, and stereotyping can be particularly difficult for minority professionals to deal with when they arise in worker–client relationships. That said, it is also important to recognize that a student of any background can experience biased attitudes from their clients based on assumptions and stereotypes. Given the complexities of the dynamics in relationships involving diversity, it has become clear that an understanding of culture alone is not sufficient to address these complexities. Consideration of "social location" can help to flesh out students' understanding of cross-cultural helping relationships. Worker–client relationships may represent various dyads of diverse cultural pairings, resulting in complicated dynamics related to power, privilege, and discrimination. For example, an African-American professional might work with Asian-American clients, or an American Indian worker might work with Latino clients. Given the complexity and diversity of the real world situations that these and other scenarios represent, the "analysis of social location" approach has been suggested as an important component of diversity education in human services, augmenting the traditional approach of introducing students to information about various cultures. Rather than assuming "whiteness," the analysis of social location approach calls upon students to identify their own social location as well as the social locations of their clients.

Analysis of social location involves examining your social position to identify issues of privilege and oppression in your own unique situation. Social location analysis assumes that to some extent imbalances of power exist within virtually all relationships and that any given individual may have both issues of power and oppression in her or his unique social location (Carniol, 2005). For example, as a White female I have experienced the privileges of "whiteness" in American society. Simultaneously, I have experienced the "ism" of my femaleness as I have at times experienced gender-based stereotypes, assumptions, and discrimination. Further, as I grew up in the rural south, I have at times experienced disadvantages, stereotypes, and negative assumptions based on the region that I come from. An African-American colleague who grew up in a family with two professional parents, a relatively high income, and upper middle class socioeconomic status also experienced a mixed social location, including both elements of privilege (related to income and socioeconomic status) and cultural oppression (related to race).

Factors in social location that create disadvantages for people are sometimes referred to as *isms*, such as sexism, racism, ageism, and heterosexism (Carniol, 2005). No discussion of culture and social location is complete without consideration of prejudice and discrimination and their potential impact on the clients you serve. Many clients have experienced long histories of prejudice and discrimination due to their race, ethnicity, cultural identity, socioeconomic class, disabilities, gender, or sexual orientation, for example. In fact, many human service clients are dealing with multiple isms that subject them to repeated forms of discrimination and few if any bases of privilege that might grant them advantages in our society.

For example, the lesbian woman who has disabilities will likely struggle with multiple isms as she encounters sexism, heteroism, and able-bodiedism in society. Yet, if she has a good education and higher socioeconomic class, these attributes will grant her certain privileges. Through these privileges, she will have far greater possibilities for a satisfying position in society than would a similar individual without these particular advantages. The point is that each of us and each of our clients have unique and complex social locations and histories. Working within these relationships with sensitivity and awareness requires the professional to be attuned to all of these various dimensions of social location, including issues of color, ethnicity, social class, educational level, gender, disability, sexual orientation, educational level, and regional identity.

Experiencing repeated discrimination can result in a range of emotional reactions in clients that you might observe—mistrust, anger, suspicion, hostility, and resentment. As you develop relationships with clients, it may be appropriate at times to talk with them about their experiences and whether they feel they have experienced discrimination and prejudice. Broaching this topic can be an important part of understanding their cultural experience and conveys your sincere interest in and empathy for their situation. Due to the intensely personal and painful nature of this discussion, it is generally important to initiate it within the context of an established supportive relationship. Bringing this sensitive topic up for discussion before trust is established could be counterproductive.

 ◆ EXERCISE 7.6 SYNTHESIS: LINKING KNOWLEDGE AND EXPERIENCE

What is your own unique social location? What elements of who you are have placed you in the position to experience discrimination or stereotyping (the " isms") that are present and active in society, such as racism and sexism? Consider issues of color, ethnicity, social class, educational level, gender, disability, sexual orientation, and regional identity, as well as any other factors that might play a role in your social location.

As you consider your unique social location, how do the privileges and challenges presented by your social location impact your work with clients? What particular sensitivities and insight do you see yourself as possessing due to your social location? In what ways does your social location enhance your ability to feel empathy for others?

Understanding Power Differentials

The social locations of people in various dyadic relationships sometimes operate to suggest a "better than" or "less than" status of one individual in relation to another. These dynamics can come into play in helping relationships (Petrie, 2004). Typically, the human service professional's social location tends to be more favorable than that of the client. The professional often has more education and a higher socioeconomic status than many clients. Less frequently, however, the professional may be in the less advantaged position, such as working with a wealthy client, a more educated client, or a person who has high status and power in the local community. In such relationships, professionals may find it more difficult to establish credibility with the client or may find that the client is treating them in a condescending manner. These situations can be challenging for even highly experienced professionals. Unequal power dynamics of either type in worker–client relationships can create challenges in the helping process that should be discussed in supervision.

In addition to interpersonal dynamics, agency policies as well as elements of agency culture (discussed in the next section) can further serve to place clients in an unempowered position. The human service literature is replete with information regarding the importance of client empowerment. Therefore as a professional, you must strive to minimize these power differentials, not only in your client relationships but also in your agency's policies and practices, in order to empower clients.

Power differentials between professionals and clients can be emphasized in a number of subtle and not-so subtle ways. One such practice, for example, may be calling clients by their first names, such as "Anna" or "Luis," but expecting clients to call professionals by more formal names, such as "Ms. Harris" or "Mr. Kim." A waiting room that is ill-furnished, uncomfortable, dirty, or ill-kempt can also convey a lack of respect for clients, indirectly sending a message that clients are not important or that their wishes are not to be considered.

Still other practices may more directly disempower clients. For example, agency policies and professional practices that do not recognize client strengths and abilities or reflect a negative perception of client capabilities clearly work against empowerment. Policies and practices that limit client choice and decision making in arbitrary ways also work against client empowerment. Furthermore, environments that are dismissive of clients' perceptions or opinions are innately disempowering.

Professionals must be sensitive to elements in agency culture that magnify the power differential between workers and clients, and they must also be willing to confront and address these issues as needed. In fact, the Ethical Standards of Human Service Professionals requires professionals to engage in "constructive criticism of the profession" (Ethical Standards, 1996). Nevertheless, student interns who observe such situations in their field settings can find it difficult to address these issues as a newcomer and a relative outsider in the organization. Like clients, interns generally experience a lack of power and privilege in relation to the more experienced, established professionals with whom they work. Confronting agency practices that need to be changed is typically more effectively done by agency insiders.

Established workers within the organization who have "earned the right to be heard" have generally been in the organization for some time and have won the respect of other staff. These are the staff members who are most likely able to effect change within the organization. In some cases, workers who have been in the agency for a longer period might be desensitized to counterproductive practices and may not recognize them, or assume leadership in changing them. As a relative newcomer, seeing the situation with fresh eyes, you may be more sensitive to issues of power and privilege as they play out in the organization. Tactfully talking with your supervisor about your observations and/or with your faculty liaison gives you the opportunity to receive feedback about your perceptions and may even open avenues for discussion and potential change within the organization. More important, monitoring your own behavior for such expressions of power is one of the most important things that you can do as a professional.

A Student's Experience with Issues of Power and Privilege in the Organization

Today I felt very sad for a little third-grade boy who was sitting by himself in the school cafeteria not eating. One of the school counselors (not my supervisor) went over and asked him why he wasn't eating. He told her that it was Ramadan and that he wasn't allowed to eat until sundown. She said to him, "That's terrible. Aren't you hungry?" The little boy said, "No." As we walked away the counselor was pretty upset, saying, "What kind of parents would make their child do without food all day? I ought to call social services!" As various teachers and counselors came into the cafeteria, she told each one about what the student had said to her and wondered aloud what they should do about it. The other staff members seemed to agree with her. I could sense that the student knew that these adults were talking about him because he kept glancing back and forth between them and the tabletop as they kept talking. I finally told my supervisor that I would like to go sit with him and keep him company, and she said that would be fine. I sat down at his table, and we chatted for a while. After a few minutes I told him that I would like to know more about Ramadan, and he told me about the traditions and rituals around this event. I was positive, encouraging, and genuinely interested in learning more about his family's traditions within Islam. As we left the cafeteria, my supervisor commented on how the student had "perked up considerably" as I had talked with him.

◆ EXERCISE 7.7 SYNTHESIS: LINKING KNOWLEDGE AND EXPERIENCE

Power differentials between professionals and clients are to some extent inevitable. How do you see these issues being played out in your internship agency or in other agencies that you might have visited during your internship? Have you observed any particular agency policies or worker behaviors that tend to accentuate these power differentials? What suggestions do you have as to how this organization might deemphasize or reduce these power differentials?

Considering the Isms and You

Unfortunately, isms and power differentials can play out in internship experiences in ways that more directly and personally involve you. Depending upon your own social location, you may have encountered prejudice and/or discrimination throughout your life. Bringing this history into your internship can make the situation particularly challenging for you in many ways. For example, if you are a woman who has been discriminated against in previous work environments, you may have difficulty trusting authority figures at your internship site. Similarly, you may have difficulty working with clients who seem to represent the power structure that discriminated against you, such as White males. If you encounter these reactions in yourself, talk with your supervisor and faculty liaison about your feelings. There may be issues and feelings that you need to confront and work through in order to gain greater comfort in these relationships.

It is also the case that you may directly experience some of the isms during your internship. As you entered your internship, you entered a human service organization that is part of the larger society with all of its prejudices and biases. Although we might hope that human service agencies would be immune to such problems, this is not always the case. Therefore, students can sometimes be subjected to bias, discrimination, or prejudice of various kinds during their internships. Discrimination of various types occurs to people on the job, and internships are no exception. In fact, interns in some settings may be particularly vulnerable to discrimination because they may be seen as relatively powerless people within the system.

Sexual harassment in internships is not unknown. Daugherty, Baldwin, and Rowley (1998), in a national survey regarding medical internships, found that 63 percent of respondents had experienced at least one incident of sexual harassment or discrimination during their medical internships. Sexual harassment may be defined as any unwelcome sexual advances, requests for sexual favors, or other verbal or physical behavior of a sexual nature that creates an intimidating, hostile, or offensive work environment. Should you experience this or other kinds of discrimination in your internship agency, you should be reassured by the fact that most agencies have clear policies against such behaviors. If you feel that you are experiencing biased reactions from clients or co-workers in your internship, discuss your perceptions and feelings with your supervisor. Through supervision, you can determine some strategies you might use to deal with the situation. The student reflection below is an example of one such internship experience, and reflects the frustration the student was feeling.

A Student's Reflections on Bias and Prejudice

Sometimes I can tell that clients are surprised to see me in the role of someone that's supposed to help them. Here I am a young African-American male, and I'm trying to help people who are suspicious of me or even downright scared of me.

continued

Some clients just assume I'm another client. My supervisor asked me if I'd gotten any racist reactions from the clients, and I was glad she brought it up. I was afraid if I brought it up myself she would think I was paranoid or something. We talked about some ways I could deal with it. She told me about when she was just starting out in the field how she had to dress more formally than she does now just so people would take her seriously. Also we talked about some things I can do if I sense that somebody is particularly uncomfortable around me, such as being especially warm and friendly toward them, taking the relationship slow, etc. It's also ok to check out with people how they feel about working with me and directly bringing up the issue of race if I feel it's an obstacle. That may be uncomfortable, but sometimes it's the best thing to do. Anybody that says racism doesn't exist anymore just needs to be me for a day. It is so ridiculous for anybody to be scared of me that sometimes it's just funny to see people react to me as just "a big black guy" instead of as "Anthony—human services major," but I have to admit that sometimes it just makes me mad and sad and very tired. It really gets old and I, of course, start getting defensive. Sometimes I just want to say, "So what's your problem?" But I don't think that would count as "professional behavior."

Understanding Agency Culture

As you explore concepts related to culture in your internship, the organization in which you work is important to consider. Organizational culture can be defined as "an organization's beliefs, knowledge, attitudes, and customs. Culture may result in part from senior managers' beliefs, but it also results from employees' beliefs. It can be supportive or unsupportive and positive or negative. It can affect employees' ability or willingness to adapt or perform well" (Bassi, 1997). Just as human services professionals must work effectively with clients from unique and distinctive cultures, they must also work within organizations with unique and distinctive cultures.

Professionals can sometimes experience culture clash as they take their own culturally learned habits and assumptions into environments where other cultural norms are dominant. In fact, entering your field agency as an intern might have presented a similar situation for you. Agencies and professions build cultures over time that can seem quite foreign to a newcomer. The term *organizational culture* recognizes that businesses and organizations function as distinct cultural entities, having their own clearly established systems of values, norms, assumptions, beliefs, expectations, behaviors, rituals, and so on (Holland, 1995; Weiner, 1990). Some agencies, for example, place a very high value on staff cohesion and camaraderie within the group, whereas others place a high value on individual effort and autonomy. Some agencies have very specific expectations regarding how employees should dress, whereas others have a far more laissez-faire attitude regarding dress. How staff members relate to each other, how they relate to clients, whether doors tend to be kept open or closed, whether lunch is a shared time within the group or is a time for people to go their separate ways—all of these are examples of how agency culture expresses itself.

You entered your agency with a unique cultural identity of your own. In addition to the cultural heritage that you gained in your family and community, as a student you have also adapted to a particular campus or student culture. Given this mix of cultural realities, many students experience their first weeks in their field placement as a process similar to adapting within a new culture.

◆ EXERCISE 7.8 PERSONAL REFLECTION: OBSERVATION OF SELF AND OTHERS

Review the first days in your field placement. What did you observe about the agency's culture? (Review the components of agency culture discussed earlier.) How did the expectations of the agency culture differ from your usual experience in the "student culture" of your college or university? The culture of your family and community? In what ways, if any, did you experience a clash between your behavior and the norms and expectations of the agency culture?

Although your agency culture might not have been drastically different from your personal culture or your campus culture, it is likely that you were able to identify at least at few areas in which you observed differences and possibly had to adapt your behavior. The clients with whom we work are often faced with similar, but often much more pronounced, culture clashes in their workplaces and in their interactions with human service organizations. An important skill in developing cross-cultural competence is gaining the ability to help clients navigate the cultural challenges that they face in our increasingly diverse world.

A Student's Reflections on Agency Culture and Student Culture

At first glance I didn't think there would be that much difference between life as a student and life in a human services agency, but by the end of the first week it was clear to me that there were some differences. The people here might look like me and my college professors, but the expectations were certainly different than on campus. First of all, because everyone had been so nice and friendly, I didn't realize that in some ways things would be so much more strict or formal. As with most things, I learned this the hard way. On the second day of my internship, we had a case conference that lasted for about one-and-a-half hours. After an hour or so, I got up and went to the bathroom and came back in. This would have been OK in most of my college classes, but after the meeting my supervisor explained to me that it wasn't "the norm" here. People stay in a meeting until it is finished unless they are called away for something really important, like an emergency.

Some other things that I have come up against have been about things like phone manner and skirt lengths. I have noticed that my natural way of asking a person on the phone to wait is to say, "Hang on a minute." Nobody even has to tell me how awful this sounds or that it does not "fit" with this agency's way of doing things. When it pops out of my mouth, it sounds horrible there, but it sounds fine to me at home. I am trying to develop a new habit of saying, "Would you please hold?" Also, recently a student volunteer came to work wearing a skirt that was too short. The volunteer coordinator asked her to go back home and change. You would see hundreds of skirts like it every day on campus. In fact, I have a skirt pretty much like it in my own closet and probably would have worn it to work eventually. At least on this one I got to learn from someone else's mistake! I've never worked in a professional environment before, and even though it doesn't bowl you over at first, there really does seem to be such a thing as a "professional culture." It is clear to me that I need to try to conform to this culture if I want to fit in.

Conclusion

Over the course of your career, issues of diversity are likely to become more pronounced in your work and in society in general. The principles discussed in this chapter will be helpful to you in working with this increasing diversity. The greatest barriers for human services workers as they form relationships with clients from different cultures tend to be a lack of cultural self-awareness, a lack of knowledge about other cultures, and ethnocentric reactions to cultural differences. Also, issues of power and privilege can interact with diversity, making these relationships even more complex.

Therefore as a student, it is important that you learn as much as possible about the cultures of the various clients with whom you work and strive to become more aware of your own cultural make-up and reactions. Monitoring your reactions to various clients and situations can help you identify feelings of rejection or judgmental attitudes that might possibly be red flags indicating cultural bias or ethnocentrism that you can work to overcome. Once you have developed cultural competence, the cultural diversity that you experience in your work can be a real source of pleasure, adding interest, depth, and variety to your professional life.

You can learn more about cultural issues in human services through numerous websites. Particularly useful is a lengthy and detailed U.S. Department of Health and Human Services report in which National Standards for Culturally and Linguistically Appropriate Service in Health Care are discussed. The principles included in this report are pertinent to all types of human service settings, not just health care. The National Center for Cultural Competence at Georgetown University also has an excellent website that focuses on the center's mission "to increase the capacity of health and mental health programs to design, implement, and evaluate culturally and linguistically competent service delivery systems" (National Center, 2004). This site includes basic information about cultural competence as well as a number of self-assessment tools related to cultural competence for both organizations and individuals.

FOR YOUR E-PORTFOLIO

What cultural differences have you encountered in your internship? Through your internship, what insights have you gained about your own culture and social location? What insights have you gained into the cultures and social locations of the clients with whom you have worked? What have been your greatest cultural challenges and how have you dealt with them? How do you assess your current level of skill in working with people who are culturally different from you? What are your goals as you continue to develop your skills in working with diverse groups?

References

Axelson, J. (1993). *Counseling and development in a multicultural society.* Pacific Grove, CA: Brooks/Cole.

Axelson, J. (1999). *Counseling and development in a multicultural society* (3rd ed.). Pacific Grove: Brooks/Cole.

Bassi, L. (1997). Harnessing the power of intellectual capital. *Training & Development*, December 1997. Retrieved November 13, 2005, from American Society of Training and Development, http://www.astd.org/astd/Resources/performance_improvement_community/Glossary.htm.

Bricker-Jenkins, M., & Lockett, P. (1995). Women: Direct practice. In R. Edwards & J. Hopps (Eds.), *Encyclopedia of social work* (pp. 2529–2539). Washington, DC: NASW Press.

Carniol, B. (2005). Analysis of social location and change: Practice implications. In S. Hick, J. Fook, & R. Pozzuto (Eds.), *Social work: A critical turn* (pp. 153–165). Toronto: Thompson.

Chau, K. (1991). Introduction: Facilitating bicultural development and intercultural skills in ethnically heterogeneous groups. In K. Chau (Ed.), *Ethnicity and biculturalism: Emerging perspectives of social group work* (pp. 1–5). New York: Haworth Press.

Corey, G., Corey, M., & Callanan, P. (2003). *Issues and ethics in the helping professions* (6th ed.). Pacific Grove, CA: Brooks/Cole.

Cox, C., & Ephross, P. (1998). *Ethnicity and social work practice.* New York: Oxford University Press.

Daugherty, S., Baldwin, D., & Rowley, B. (1998). Learning satisfaction, and mistreatment during medical internships: A national survey of working conditions. *Journal of the American Medical Association, 279*(15), 1194–1199.

Day, J. (2000). U.S. Bureau of Census: National Population Projections. Retrieved November, 13, 2005, from, http://www.census.gov/population/www/pop-profile/natproj.html.

de Anda, D. (1984). Bicultural socialization: Factors affecting the minority experience. *Social Work, 29*, 101–107.

De Vos, G. (1975). Ethnic pluralism: Conflict and accommodation. In G. De Vos & L. Romanucci-Ross (Eds.), *Ethnic identity: Cultural continuities and change* (pp. 5–41). Palo Alto, CA: Mayfield.

Ethical Standards of Human Service Professionals. (1996). *Human Service Education, 16*(1), 11–17.

Ewalt, P., & Mokuau, N. (1995). Self-determination from a Pacific perspective. *Social Work, 40*, 168–175.

Gilligan, C. (1982). *In a different voice.* Cambridge, MA: Harvard University Press.

Halley, A., Kopp, J., & Austin, M. (1998). *Delivering human services: A learning approach to practice.* New York: Longman.

Holland, T. (1995). Organizations: Contexts for social services delivery. In R. Edwards & J. Hopps (Eds.), *Encyclopedia of social work* (pp. 1787–1794). Washington, DC: NASW Press.

Kendall, D. (2006). *Sociology in our times* (5th ed.). Belmont, CA: Wadsworth.

Lee, C. (1991). New approaches to diversity: Implications for multicultural counselor training and research. In C. Lee & B. Richardson (Eds.), *Multicultural issues in counseling: New approaches to diversity* (pp. 3–9). Alexandria, VA: American Association for Counseling and Development.

Longres, J. (1995). Hispanics overview. In R. Edwards & J. Hopps (Eds.), *Encyclopedia of social work* (pp. 1214–1221). Washington, DC: NASW Press.

Lukes, C., & Land, H. (1990). Biculturality and homosexuality. *Social Work, 35*, 155–161.

Lum, D. (1999). *Culturally competent practice.* Pacific Grove, CA: Brooks/Cole.

Maslow, A. (1954). *Motivation and personality.* New York: Harper and Row.

National Center for Cultural Competence. (2004). Georgetown University Center for Child and Human Development. Retrieved November 17, 2005, from http://gucchd.georgetown .edu//nccc/.

Nerney, B. (1998). Preparing interns to discover common ground. *NSEE Quarterly, 23*(4), 1, 16–22.

Neukrug, E. (2004). *Theory, practice, and trends in human services* (3rd ed.). Pacific Grove, CA: Brooks/Cole.

Petrie, R. (1999). Trends and challenges of cultural diversity. In H. Harris & D. Maloney (Eds.), *Human services: Contemporary issues and trends* (2nd ed., pp. 393–399). Boston: Allyn & Bacon.

Petrie, R. (2004). Trends and challenges of cultural diversity. In H. Harris, D. Maloney, & F. Rother (Eds.), *Human services: Contemporary issues and trends* (3rd ed., pp. 359–366). Boston: Allyn & Bacon.

Robinson, T. (2005). *The convergence of race, ethnicity, and gender: Multiple identities in counseling.* Upper Saddle River, NJ: Pearson.

Schlesinger, E., & Devore, W. (1995). Ethnic-sensitive practice. In R. Edwards & J. Hopps (Eds.), *Encyclopedia of social work* (pp. 902–908). Washington, DC: NASW Press.

Schram, B., & Mandell, D. (1994). *An introduction to human services: Policy and practice.* New York: Macmillan.

Sue, D., Ivey, A., & Pedersen, P. (1996). *A theory of multicultural counseling and therapy.* Pacific Grove: Brooks/Cole.

Valentine, C. (1971). Deficit, difference, and bicultural models of Afro-American behavior. *Harvard Educational Review, 41*, 137–159.

Van Den Bergh, N. (1991). Managing biculturalism at the workplace: A group approach. In K. Chau (Ed.), *Ethnicity and biculturalism: Emerging perspectives of social group work* (pp. 71–101). New York: Haworth Press.

Weaver, H. (1998). Indigenous people in a multicultural society: Unique issues for human services. *Social Work, 43*, 203–211.

Weiner, M. (1990). *Human services management: Analysis and applications* (2nd ed.). Belmont, CA: Wadsworth.

Yee, J. (2005). Critical anti-racism praxis: The concept of whiteness implicated. In S., Hick, J., Fook, & R. Pozzuto, (Eds.), *Social work: A critical turn.* Toronto: University of Toronto. (Canadian)

Chapter 8

Developing Ethical Competence

In recent years there has been intense interest in ethical issues within most professions because issues of legal liability have become prevalent in society in general. A number of factors are thought to have contributed to this increased interest in professional ethics and legal liability. Technological advances that have created new ethical dilemmas, difficult decisions regarding the allocation of scarce resources, and widely publicized ethical scandals and litigation within many professions have fueled the interest of professionals and laypeople alike in ethical practice (Reamer, 1995).

Throughout your human service program, professional ethics has probably been strongly emphasized, and rightly so. Many authors have written at length about the significance of ethics in human service work. For example, Cournoyer writes, "Ethical responsibilities take precedence over theoretical knowledge, research findings, practice wisdom, agency policies, and, of course, your own personal values, preferences, and beliefs. . . . Ethical decision making is an essential component of professionalism" (2005, p. 82).

Ethical decision making is at the center of all that you will do in your fieldwork, all that you observe others do, and all that you do in your career as a human service professional. Your fieldwork experience offers an ideal opportunity to develop and fine-tune your ethical skills. If you are not familiar with the *Ethical Standards of Human Service Professionals,* you should read it carefully. The full Ethical Standards document can be found in the Appendix of this book and on the website of the National Organization for Human Services.

An Overview

Throughout each day, ethical issues are identifiable in every situation that you experience in the field. In many situations, however, you might not think about the ethical issues involved because the situation is so straightforward. For example, issues of privacy and confidentiality are involved whenever you close the door, talk with a client, and carefully handle the information shared. Issues of client respect are involved whenever you express acceptance rather than judgment toward a client. In these situations, your behavior is being directed by the ethics and values of the profession, although you may not be thinking consciously about this fact. At other times, the ethical issues in a situation take center stage, becoming painfully obvious because they are complex and not easily resolved. For example, at what point do you break confidentiality to report suspected child abuse? What do you do when you are angered or repulsed by a client's behavior? Your thoughts in such situations inevitably turn to ethical considerations as you search for a proper direction to take.

Although some situations are inherently more ethically challenging than others, ethical action in all situations calls upon the professional

1. to be knowledgeable about the legal and ethical foundations of the human service profession.
2. to understand the values and ideals that guide responsible human services practice.
3. to be able to recognize the specific ethical, legal, and values issues involved in any given practice situation.

These skills are required in even the most straightforward situations in which there is not a values conflict or an ethical dilemma. When such conflicts and dilemmas do exist, the following additional skill is needed:

4. to have the ability to weigh the conflicting ethical and values positions involved in a given situation and make decisions between them.

These four skills might be thought of as ethical competencies, and each is addressed in this chapter.

Competency 1: Understanding the Ethical and Legal Foundations of the Profession

The foundation for ethical action rests in part on knowledge and understanding of the human service ethical standards. (See the Appendix for the complete statement of *Ethical Standards of Human Service Professionals*.) Certainly all practitioners should consult these standards when confronted with a difficult ethical situation. In many cases, the standards are clear in determining the action or stance of the professional. In such cases, no further decision making is necessary (Welfel, 2006). In this section, you can read a summary of the *Ethical Standards*. As this summary reflects, the ethical standards are broad and far-reaching, addressing professional behavior in a wide range of areas. In addition to professional ethics, this section also deals with the legal standards that commonly affect practice in human service settings.

Ethical Standards of the Human Service Profession

Read the following summary of the standards with your field agency in mind. As you read, reflect upon situations that you have encountered during your fieldwork, as a participant or as an observer, in which one or more of these ethical issues came into play.

◆ *The human service professional's responsibility to clients.* Clients have the right to information about the helping process, the right to receive or refuse services, the right to be treated with respect and dignity, the right to confidentiality, and the right to know about the limits of confidentiality. Workers are obligated to avoid dual worker–client relationships, to protect clients and others from harm, to maintain secure records, and to recognize client strengths.

◆ *The human service professional's responsibility to the community and society.* Professionals adhere to state, local, and federal laws and advocate for changes in those laws that conflict with clients' rights. Professionals stay informed on social issues affecting clients, engage in advocacy and mobilizing to satisfy unmet human needs, and advocate for the human rights of all people. Human services professionals are knowledgeable about their own culture and cultures within their community, respect cultural and individual differences, and seek training and expertise in working with culturally diverse populations. Workers act to protect the safety of others who might be harmed by clients' behavior and represent their own qualifications and services honestly.

◆ *The human service professional's responsibility to colleagues.* Professionals treat colleagues with respect, dealing with any conflicts directly with that colleague in effort to resolve the problem. Applying a similar principle, workers report ethical violations of colleagues, initially speaking to the colleague directly in an effort to resolve the problem. Workers avoid duplicating the efforts and relationship of another professional in working with a client and consult with other professionals as necessary to benefit the client. Consultations between professionals are kept confidential.

- *The human service professional's responsibility to the profession.* Professionals act with integrity, recognizing the limits of their expertise, seeking consultation, and making referrals as necessary. They continue to seek professional growth and learning for themselves throughout their careers as well as promote the continued development of the profession. Professionals promote cooperation among the various disciplines within human services.
- *The human service professional's responsibility to employers.* Professionals keep their commitments to their employer and participate in efforts to maintain positive working conditions and evaluate agency effectiveness.
- *The human service professional's responsibility to self.* Professionals strive to maintain personal qualities that are associated with effectiveness in the profession, foster their own self-awareness and personal growth, and commit themselves to ongoing learning and skill development (*Ethical Standards of Human Service Professionals,*1996, pp. 11–17).

In view of the wide scope and rigor of these standards, Blair's statement regarding ethical practice is understandable when she says, "It would be unlikely that any human service worker could get through a whole month of working in the field without being able to provide some example of behavior observed that they felt was unethical, or perhaps breaking a law" (1996, p. 187). Human service work deals inevitably with important values and competing "shoulds" as well as real limitations on what can be done. For example, workers and agencies are expected to act in support of client well-being at all times but in reality often work with such limited resources that the quality of services suffers. Likewise, professionals are rightly expected to be empathic, nonjudgmental, and nonmanipulative toward their clients as well as direct and straightforward in addressing conflicts with colleagues. Yet workers, like all human beings, are fallible with emotions and impulses that are not always perfectly contained or managed. Within this mix of ideals with reality, it is safe to say that there are no perfect agencies and no perfect workers, though all must strive to maintain the highest possible ethical standards.

In addition to a solid knowledge of the ethical standards, it is wise to give some thought to which particular standards you are at greatest risk of violating within your own setting and within your own work. Each agency, depending upon the nature of the work it does as well as its own inherent strengths and weaknesses, has its potential pitfalls. For example, in some agencies the intense nature of the demands upon the staff might increase the likelihood that conflict will develop between staff members; therefore, the standards having to do with colleague relationships might be particularly at risk. In another agency, tough decisions about confidentiality may be predictably at risk due to the population being served. For example, when working with potentially violent clients, issues regarding "duty to warn others" are likely to surface.

Legal Issues and Standards of the Profession

Related to, but not identical with, ethical standards are legal issues that must also guide the decisions and actions of human service professionals. Although professional organizations monitor and sanction their members' behavior in relation to the ethical standards of the profession, legal obligations carry even more weight. In the case of legal issues, the broader society sets certain expectations, duties, and obligations for professionals. Unfortunately, a clear, concise statement of these expectations does not exist. Legal standards change and evolve over time, emerging from court decisions, legislation, and regulations. As a result of this dynamic and evolutionary quality, the legal standards impacting workers may not be completely clear because they are continuously being shaped (Cournoyer, 2005). Professionals should be aware, nonetheless, that they work in an environment of not only ethical constraints but legal constraints as well. It is generally agreed that human services workers have the following legal obligations:

◆ to provide a reasonable standard of care,
◆ to respect client privacy and maintain confidentiality,
◆ to inform clients accurately and thoroughly about the services offered,
◆ to report abuse, neglect, or exploitation of dependent people, and
◆ to warn and attempt to protect people who might be harmed by a client's behavior. (Alle-Corliss & Alle-Corliss, 1998, 1999; Cournoyer, 2005; Everstein & Everstein, 1983)

Most of these legal standards are also addressed in the professional ethical standards, but the one area that is not directly addressed is that of "reasonable standard of care." The ethical standards do speak to issues of quality of care throughout in a general way, often identifying ideal standards of worker behavior. Although professionals and agencies should strive toward the highest ideals in their conduct, unfortunately legal questions about standards of care often have to do with the lower limits of acceptable worker and agency performance, especially when litigation against professionals and agencies is involved. Standard of care is perhaps the broadest of the legal issues, and one which clearly relates to each and every human services setting. It is morally and ethically right, as well as legally required, that all professionals and agencies provide a reasonable standard of care to the people whom they serve. But what exactly is a reasonable standard of care?

A reasonable standard of care relates to many areas of worker activity but generally is thought to include at least the following:

◆ knowledge about the population being served and the services being offered,
◆ delivery of services and interventions based on sound theoretical principles,
◆ reliability and availability of services to clients,
◆ initiative and action on behalf of client and public safety,
◆ adherence to ethical standards of the profession in relation to client care, and
◆ systematic, accurate, thorough, and timely documentation of client care.

◆ EXERCISE 8.1 SYNTHESIS: LINKING KNOWLEDGE AND EXPERIENCE

Think about your own fieldwork site. What particular *ethical* standards are dealt with frequently within this setting? Identify the specific standards in question and explain why these standards are particularly pertinent and challenging within your setting.

Also identify the legal issues and standards that are likely to be most relevant to your setting. What circumstances give these issues prominence in your setting?

As charges of malpractice become more common, human service professionals and the agencies in which they work must stay abreast of the legal issues that might emerge in their work. Malpractice charges may allege either professional misconduct or negligence on the part of a professional or agency. Therefore, acts of commission (what the professional did) as well as acts of omission (what the professional did not do) are open to such charges (Kurzman, 1995). Human service professionals, of course, can reduce their risk of becoming involved in such charges by staying aware of the legal and ethical risk factors that are potentially present in their work and by consistently maintaining the highest standards of practice.

A Student's Reflections on Ethical and Values Issues

I know this isn't illegal but I certainly consider it unethical. There are a few workers here who talk to the children in a really hateful, angry tone of voice most of the time. I can honestly say that I have never heard two workers in particular say a nice word to the children. It's not my place to be correcting the paid staff members, so I just bite my tongue. I can't believe that they would talk to adults this way. I believe that the human service values of treating people with respect, dignity, and warmth should apply to children, as much as they do to adults. I plan to ask my supervisor about this, but I don't want to seem critical. In the workers' defense, I have noticed that there is no training for new staff coming into the organization and very little in writing about how the staff should relate to the children. I believe that the workers at times fail the children by not giving them what they deserve, but I also believe the agency may be failing the workers as well by not giving them the training and support they need. Whatever the explanation for the problem, I believe that a higher level of care could be and should be provided for these children.

Competency 2: Understanding the Values and Ideals That Should Guide Responsible Human Services Practice

In addition to having a knowledge of the ethical and legal standards of the profession, workers who strive to act ethically must also develop an understanding of the values and ideals of the profession that shape the larger context of ethical action. Certain worker characteristics may be thought of as professional values within the human service profession because there is a high degree of consensus among members of the profession that these characteristics are desirable. Some of the most commonly cited professional values are human dignity, respect, client self-determination, empathy, genuineness, positive regard, nonpossessive warmth, patience, self-awareness, pragmatism, noncontrolling interpersonal style, trustworthiness, and open-mindedness (Egan, 1994; Mehr, 1998; Neukrug, 2004; Schram & Mandell, 2006; Woodside & McClam, 2006). While some of these values are specifically mentioned within the *Ethical Standards of Human Service Professionals*, some are not. Yet the literature within the field clearly sets a context of expectations for competent professional behavior by consistently citing the traits previously noted, among others, as desirable. Each of these values potentially has direct bearing on the moment-by-moment choices made by human service professionals as they consider such questions as: "How will I respond to this individual?" "What action will I take in this situation?" "What must I be careful not to do or say at this point?"

 ◆ **EXERCISE 8.2 PERSONAL REFLECTION: OBSERVATION OF SELF AND OTHERS**

Review the values listed previously as desirable worker traits. Expand upon this list by identifying additional commonly cited professional values not listed above, based upon your recall of your previous coursework.

Which of the professional values do you see as generally most descriptive of you? Which of the professional values do you see as generally least descriptive of you?

Within your particular fieldwork setting, which of these values are you finding most difficult to live out in your daily decisions, behavior, and attitudes? Explain the aspects of your work that make these values particularly challenging.

Competency 3: Recognizing the Ethical, Legal, and Values Issues Involved in Any Given Practice Situation

Occasionally, interns say that they have encountered no ethical issues during their internships. What students usually mean by this is that they have not encountered any gross ethical violations. The student might have a good knowledge of professional ethics, legal standards, and values but not be fully attuned to their relevance in everyday situations. An important objective of the fieldwork experience is that students develop ethical sensitivity—that is, the ability to recognize the ethical, legal, and values issues involved in their day-to-day work. Ethical sensitivity, as compared to ethical knowledge, goes beyond just knowing the ethical standards or recognizing gross ethical violations and extends into the ability to recognize more subtle ethical, legal, and values issues that operate within ordinary, everyday practice situations.

Human services professionals are expected to be sufficiently aware of and reflective about their work to know the issues inherent in a given situation before they get pulled unknowingly into questionable behavior or circumstances. As a student, you are particularly urged to take time for careful reflection each day. Using the Integrative Processing Model (discussed in Chapter 4) will help to ensure that you are thinking through your experiences carefully enough to identify feelings of dissonance and to recognize the ethical, legal, and values issues that are present within them. For practice, it may be helpful to look at a few examples of difficult situations that other students have experienced and identify the issues involved.

CASE EXAMPLE 1

Anna was a human services intern at a crisis center for abused children. The center consisted of several components, including a group home, a victim advocacy program, and a community education program. Each program had its own coordinator who reported to the agency director. Anna's field supervisor was the agency director, Tyrone.

Tyrone suggested that Anna spend time in each of the agency's programs, shadowing and accepting assignments from each of the coordinators. This was agreeable with Anna because it conformed to her own hopes of getting a broad exposure to the full range of services provided within the agency.

Within a few days, it became clear to Anna that there was a significant amount of tension between two of the program coordinators. Julie, the shelter coordinator, complained to Anna about Jeff, the community education coordinator, and Jeff complained to Anna about Julie. Jeff enjoyed spending time at the shelter and felt that the children needed some attention from a caring male as well as a good male role model. Julie felt that Jeff was intrusive and should focus on his own job rather than spending time at the shelter. Jeff felt that Julie was overly controlling, overly possessive, and overly protective with matters concerning life at the shelter.

For the first couple of weeks, Anna listened to both Julie and Jeff, making minimal responses and trying not to get involved in their conflict. Soon, however, Anna began to experience some of the same feelings toward Julie that Jeff had described. When Anna tried to get involved with the children at the shelter, she felt that Julie often assigned her something else to do to put distance between Anna and the children. She would ask Anna to run errands or get supplies ready for an upcoming meeting instead of allowing her to interact with the children.

Anna shared with Jeff her feelings and observations about Julie's behavior toward her. Soon Anna and Jeff were talking daily about the topic, trying to think up ways to crash through what they perceived as Julie's excessive and unhealthy control of life at the shelter. One day as they were talking in the crisis center office, Julie suddenly stormed into the room, furious at both Jeff and Anna. Although they thought that Julie was at the shelter as usual, she had been doing some work in the room next door and had overheard Anna's and Jeff's remarks, criticizing and ridiculing her. She furiously reported that she would be speaking to Tyrone about the conversation she had overheard, and in the meantime she did not want either of them setting foot in the shelter.

Upset to the point of tears, Anna felt angry and misunderstood. She felt that she herself was the "victim." After all, it was she who had been assigned and had completed all of the "dirty work" from Julie, and now Julie had the nerve to attack her. On another level, however, Anna felt very uncomfortable with her own part in the problem. She knew that she had contributed to the problem and should have made some different choices about her own behavior along the way. Unfortunately, Anna had not recognized previously that there were ethical issues involved in this situation. Her awareness of this came only after her unpleasant confrontation with Julie prompted extensive reflection.

◆ EXERCISE 8.3 ANALYSIS

What ethical, legal, or value issues were involved in this situation?

What might have happened differently if Anna had recognized the issues in this situation and her own obligations earlier?

What options were open to Anna early in the placement when she first began to experience problems with Julie? Which option(s) would have been most desirable for her to implement?

What options are available to Anna at this point, following the confrontation? Which option(s) would you select if you were in Anna's position?

CASE EXAMPLE 2

David, a senior human service major, was excited about his placement in the local probation and parole office for his internship. He had the opportunity to work with and observe several staff members and was then given a small caseload of his own to monitor and work with. His faculty supervisor noticed in reading David's journal that there was a great deal of variation in his communication with his clients. Some clients he spoke with often and conducted professional helping interviews involving important exploration, decision making, and goal setting. Other clients he spoke with rarely and did little more than lecture and give commands. The faculty member spoke with

David about this during a site visit, and David responded with the following comment, "You see, one thing that I've learned in this field placement is that for some of these people, it is not even worth trying. They don't care about their own life. They're not making any effort, so why should the probation officer make an effort? You can just tell which ones are worth the trouble and which ones aren't. I've talked with several officers about it, and they all agree. You're just wasting your time trying to help a lot of these people."

 ◆ **EXERCISE 8.4 ANALYSIS**

David does not seem to be aware of the professional issues involved in this situation. What ethical, legal, or values issues are present in David's situation?

If you had the opportunity to respond to David, what might you say?

Assuming that David has acquired his attitudes and behaviors from professional role models within the agency, how might he address these issues day to day in his work as an ethically sensitive helper?

CASE EXAMPLE 3

Keisha, a human services student, was performing her human services practicum at a child development center in a relatively affluent part of town. The center served mostly middle-class, professional, or two-career couples but also served a small group of children whose care was paid for by the Department of Social Services (DSS) as part of the workfare program. As Keisha started her fieldwork, the center was embroiled in a child abuse and neglect allegation involving one of the DSS children. Staff members had repeatedly noticed marks on the child that resembled a belt buckle. The decision was made to report their observations to DSS. The staff explained to Keisha the laws that required them to report suspected child abuse or neglect. They emphasized that reporting was mandatory and that they had no choice, legally, ethically, or morally, about making the report.

Within a few weeks, Keisha noticed similar marks on another child while supervising a bathroom break. She reported her observation to the teacher in her classroom. The teacher said that she had noticed similar marks on the child before but felt sure that this was not an abusive situation because this child came from such a "wonderful family." She explained that the parents were well known in the community. The father was a successful, well-liked attorney, and the mother was enrolled in a local college finishing her degree. They lived in a beautiful neighborhood, and both were pleasant, responsible people. The subject was then dropped.

Keisha continued to see marks on the child from time to time and always reported her observations to the teacher. She became more and more concerned when nothing was done and felt very confused about what her own role should be in "going over the teacher's head" about it. After discussing the situation in field seminar, Keisha decided to talk with the teacher directly but calmly about the seriousness of her concerns. In this conversation she even brought up the contrast in how this child was being dealt with in comparison to the DSS child. Despite her frustration, she was careful not to become accusatory toward the teacher or the agency. The teacher responded well to the conversation and suggested that they meet together with the director of the center to seek some guidance on the issue. The director asked both Keisha and the teacher to begin reporting their observations directly to her. She explained that she would document and verify these reports with her own follow-up observations of the child. She explained that she felt that they could not make a report at that time because "nothing was documented."

◆ **EXERCISE 8.5 ANALYSIS**

What ethical, legal, or values issues were involved in this situation?

As a student in the agency, did Keisha's options for dealing with this situation differ from those of a staff member? Why or why not?

What is your evaluation of Keisha's handling of the situation? What are the positive aspects? What are the negative aspects?

Like the students in the examples above, all human services agencies and workers confront ethical, legal, and values issues on a daily basis. Ethically sensitive workers are aware of and are able to identify the many issues involved in everyday practice situations.

◆ EXERCISE 8.6 SYNTHESIS: LINKING KNOWLEDGE AND EXPERIENCE

Identify at least one situation that you have either dealt with directly or been aware of in your field placement in which the ethical, legal, or values issues were particularly interesting or challenging. Describe the situation clearly.

Identify the *issues* involved in the situation, categorizing them accurately as *legal*, *ethical*, and/or *values* issues.

Competency 4: Weighing Conflicting Ethical and Values Positions in a Given Situation and Making Decisions Between Them

Many practice situations involve a number of ethical, legal, and/or values issues, each potentially implying a different course of action for the worker. These are among the most challenging situations confronted by human service professionals. Some method must be used for sorting out the various issues involved and deciding which should take highest priority in the worker's decision making and interventions.

The Ethical Principles Screen

Loewenberg and Dolgoff (2000) suggest that a hierarchy exists among various ethical principles and that workers should give higher priority to some principles and lower priority to others when ethical principles compete. They refer to this hierarchy as the Ethical Principles Screen. The Ethical Principles Screen is constructed as follows*:

> *Ethical Principle 1: Principle of protection of life.* The worker's highest obligation is that of protecting human life. This principle takes precedence over all others.
>
> *Ethical Principle 2: Principle of equality and inequality.* Persons of equal status and power have the right to be treated equally while those of unequal status "have the right to be treated differently if the inequality is relevant to the issue in question" (p. 61). In other words, professional interventions should acknowledge the differential levels of status and power among the people involved. For example, a child does not have the same degree of power in a relationship as does an adult. Therefore, a professional would need to take greater care to protect the child than to protect the parent's privacy and confidentiality in the case of an abusive relationship.
>
> *Ethical Principle 3: Principle of autonomy and freedom.* The professional's decisions and interventions should respect the client's right to make his or her own choices

---

*Reproduced by permission of publisher, F. E. Peacock Publishers, Inc., Itasca, Illinois. From Frank M. Loewenberg & Ralph Dolgoff, *Ethical Decisions for Social Work Practice,* copyright 2000, pp. 60–62.

and should be directed toward developing the client's ability to function independently.

Ethical Principle 4: Principle of least harm. The professional should intervene in the manner that creates the least harm for the client.

Ethical Principle 5: Principle of quality of life. The professional should make choices that enhance the quality of life for the client and others.

Ethical Principle 6: Principle of privacy and confidentiality. The professional should respect client confidentiality and the privacy of the client's home, personal life, and information.

Ethical Principle 7: Principle of truthfulness and full disclosure. The professional should report all information fully and accurately to clients and others who are authorized to have access to such information (Loewenberg & Dolgoff, 2000, pp. 60–62).

Within this framework, each principle takes precedence over the ones listed below it in the hierarchy. In cases in which various ethical principles are in conflict, this model suggests that the worker should identify the specific ethical principles involved and make decisions in accordance with the issue that is highest in the hierarchy. Read the following case example with the Ethical Principles Screen in mind.

CASE EXAMPLE 4

Sophie is 13-year-old who smokes marijuana several times per day, shoplifts regularly, and engages in unprotected sex with multiple partners. Her parents allow her to come and go as she pleases and on many occasions give her no curfew for coming home at night. It is not clear how much Sophie's parents know about her activities. Sophie says her parents know what she has been doing, but at the same time she does not want you or anyone else to discuss these issues with them. The ethical dilemma for staff members in the agency is whether to inform Sophie's parents of her activities, and, if so, how much information should be shared.

◆ EXERCISE 8.7 ANALYSIS

Using the Ethical Principles Screen as your guide, discuss the ethical issues involved in Sophie's situation. Starting at the top of the hierarchy and working your way to the bottom, consider each ethical principle and whether it applies to this situation. After working your way through the Ethical Principles Screen, decide upon and describe a course of action that you believe would be advisable, remembering that those principles at the top of the hierarchy should carry more weight in your decision than should those toward the bottom of the hierarchy.

◆ EXERCISE 8.8 SYNTHESIS: LINKING KNOWLEDGE AND EXPERIENCE

Now apply the Ethical Principles Screen to a situation drawn from your own experience in your fieldwork. (You might wish to use a situation that you described in either Exercise 8.5 or 8.6.) Identify the issues involved in the situation as represented in the Ethical Principles Screen, working your way from the top of the hierarchy to the bottom as you did in the previous exercise. Paying attention to the placement of each of the principles relative to one another in the hierarchy, decide upon and describe a course of action that takes into account the priorities of the Ethical Principles Screen.

How effective was the Ethical Principles Screen in helping you to weigh appropriately the various issues involved in this particular situation? What were the strengths and weaknesses of this approach?

Ethical Decision Making

Although the Ethical Principles Screen is a very helpful tool in making many ethical decisions, it is not in itself a complete decision-making model. Numerous ethical decision-making models have been suggested within the human service literature (e.g., Corey, Corey, & Callanan, 2003; Gross & Capuzzi, 1999; Kitchener, 1984; Forester-Miller & Davis, 1996). Many ethical situations require the professional to make clearly thought-out decisions about how to proceed. A step-by-step decision-making model is a useful tool, enabling the worker to think systematically and carefully about the situation. Corey, Corey, and Callanan identify such a model based upon their analysis of several ethical decision-making models in the human service literature**:

1. Identify the problem or dilemma.
2. Identify the potential issues involved.
3. Review the relevant ethical guidelines.
4. Know the applicable laws and regulations.
5. Obtain consultation.
6. Consider possible courses of action.
7. Enumerate consequences of various courses of action.
8. Decide on the best course of action. (2003, p. 20–21)

These decision-making steps suggest that careful thinking about ethical issues precedes ethical action. After clarifying the various ethical and legal issues involved,

---

**From *Issues and Ethics in the Helping Professions*, 6th ed., by G. Corey, M. S. Corey, and P. Callanan, 2003. Used with permission of Wadsworth Publishing, a division of International Thomson Publishing.

you must imagine alternative scenarios and their probable outcomes before selecting a course of action. Thus, abstract considerations about ethical principles in steps 1–4 must be seen in the light of pragmatic reality in steps 6–8. An important component in accomplishing both of these tasks effectively lies in step 5, seeking consultation. While this is good practice even for the most experienced professional, it is imperative for you as a student in your internship. In all cases, you should discuss any ethical concerns you might have with your field supervisor. Discussing such situations with an experienced, respected colleague is always an invaluable method of clarifying your own thoughts, brainstorming alternatives, and gaining the perspective of a less-involved but knowledgeable person.

◆ EXERCISE 8.9 SYNTHESIS: LINKING KNOWLEDGE AND EXPERIENCE

Identify a decision involving competing ethical principles and/or values that you have either dealt with directly or at least have been aware of in your field agency. Use the ethical decision-making steps outlined above to think through this situation carefully. Write below your responses to each step of the process.

Step 1: Identify the problem or dilemma.

Step 2: Identify the potential issues involved.

Step 3: Review the relevant ethical guidelines.

Step 4: Know the applicable laws and regulations.

Step 5: Obtain consultation. (If you have discussed the situation with a colleague or colleagues, summarize these discussions. If not, identify one or two individuals whom you might approach to discuss the situation, then continue through Steps 6–8.)

Step 6: Consider possible courses of action.

Step 7: Enumerate consequences of various courses of action.

Step 8: Decide on the best course of action.

As a final step in evaluating your selected course of action, you might refer again to the Ethical Principles Screen. Is your selected course of action consistent with the hierarchy of principles reflected there? If not, you may wish to review the decision-making steps to make sure that you are comfortable with your decision.

A Student's Reflections on Ethical Decision Making

I am coming to the conclusion that most of the ethical issues in my organization have more to do with acts of omission than acts of commission. What I mean is that it is not as though the staff does bad or harmful things, but at times they just don't do what they can to be there for the clients. The social worker in skilled nursing, for example, spends no time with the residents. I know she is busy with meetings and paperwork and such, but what does the word "social" in social work mean? I feel that she could decide to mark out a couple of hours a week in her schedule just to go around and visit the residents. Until this internship, I had a different, more narrow understanding of ethical issues. Now I see that every minute of the day you are making ethical decisions just by deciding how to spend your time.

Conclusion

Learning to identify ethical issues and handle them appropriately is a professional skill that you will continue to develop throughout your career. Such issues can be extremely complex and difficult to sort out. Unfortunately, there is not always one clear "right thing to do." The best guiding principles to take away from this discussion are to: (1) maintain constant awareness of the ethical and legal standards as well as the values of the profession as you go about your daily activities, (2) work on cultivating your sensitivity to ethically questionable situations so that you can recognize them early and prevent problems, and (3) once you have identified a difficult issue, seek consultation from experienced professionals as you make decisions regarding how to proceed. As you gain more experience and practice, your ability to weigh competing principles and make ethically responsible decisions will grow.

If you would like information about the ethical statements of other human service professional groups, you will find examples on the websites for the National Association for Social Workers, the American Counseling Association, and the American Psychological Association.

FOR YOUR E-PORTFOLIO

Human service employers have a strong interest in hiring workers who have a knowledge of ethical standards, who have developed ethical sensitivity, and who can be counted on to handle ethical dilemmas in a professional manner. A common question in interviews for human services jobs is, "Tell me about a difficult ethical issue that you have dealt with and how you handled it." For your e-portfolio, write a narrative regarding a particular ethical issue that you identified and acted upon. Point out the specific actions on your part that reflect your ethical knowledge, sensitivity, and behavior.

References

Alle-Corliss, L., & Alle-Corliss, R. (1998). *Human service agencies: An orientation to fieldwork.* Pacific Grove, CA: Brooks/Cole.

Alle-Corliss, L., & Alle-Corliss, R. (1999). *Advanced practice in human service agencies: Issues, trends, and treatment perspectives.* Belmont, CA: Wadsworth.

Blair, N. (1996). Law, ethics, and the human services worker. In H. Harris & D. Maloney (Eds.), *Human services: Contemporary issues and trends* (pp. 183–193). Pacific Grove, CA: Brooks/Cole.

Corey, G., Corey, M., & Callanan, P. (2003). *Issues and ethics in the helping professions* (6th ed.). Pacific Grove, CA: Brooks/Cole.

Cournoyer, B. (2005). *The social work skills workbook* (4th ed.). Belmont, CA: Thomson Brooks/Cole.

Egan, G. (1994). *The skilled helper: A problem-management approach to helping.* Pacific Grove, CA: Brooks/Cole.

Ethical Standards of Human Service Professionals. (1996). *Human Service Education 16*(1), 11–17.

Everstein, D., & Everstein, L. (1983). *People in crisis: Strategic therapeutic interventions.* New York: Brunner/Mazel.

Forester-Miller, H., & Davis, T. (1996). *A practitioner's guide to ethical decision-making.* Alexandria, VA: American Counseling Association.

Gross, D., & Capuzzi, D. (1999). Ethical and legal issues in counseling and psychotherapy. In D. Gross & D. Capuzzi (Eds.), *Counseling and psychotherapy: Theories and interventions* (2nd ed., pp. 43–64). Englewood Cliffs, NJ: Prentice-Hall.

Kitchener, K. (1984). Initiation, critical evaluation, and ethical principles: The foundation for ethical decisions in counseling psychology. *Counseling Psychologist, 12*(3), 43–55.

Kurzman, P. (1995). Professional liability and malpractice. In R. Edwards & J. Hopps (Eds.), *Encyclopedia of social work* (pp. 1921–1927). Washington, DC: NASW Press.

Loewenberg, F. M., & Dolgoff, R. (2000). *Ethical decisions for social work practice* (6th ed.). Itasca, IL: Peacock.

Mehr, J. (1998). *Human services: Concepts and intervention strategies.* Boston: Allyn & Bacon.

Neukrug, E. (2004). *Theory, practice, and trends in human services* (3rd ed.). Pacific Grove, CA: Brooks/Cole.

Reamer, F. (1995). Ethics and values. In R. Edwards & J. Hopps (Eds.), *Encyclopedia of social work* (pp. 893–902). Washington, DC: NASW Press.

Schram, B., & Mandell, B. (2006). *An introduction to human services: Policy and practice* (6th ed.). Boston: Allyn & Bacon.

Welfel, E. (2006). *Ethics in counseling and psychotherapy: Standards, research, and emerging issues* (3rd ed). Belmont, CA: Thomson.

Woodside, M., & McClam, T. (2006). *An introduction to human services* (5th ed.). Belmont, CA: Thomson.

Young, M., & Long, L. (1998). *Counseling and therapy for couples.* Pacific Grove, CA: Brooks/Cole.

Chapter 9

Writing and Reporting Within Your Field Agency

Students choosing human services as a profession generally make this choice based upon their interest in people, human relationships, and social problems. No one enters the field because they like to do paperwork or present information before a group. Nevertheless, both types of reporting are frequently required of students in their internships and throughout their careers. As a student, whether you are in an administrative role or a direct service role, you will find that both written and oral reports of various kinds are a required part of the daily activities of human service professionals. Unlike some aspects of your internship role, this may be a task for which you have little formal preparation.

Writing, documentation, and communication of all kinds within human service settings have been transformed by the application of computer technology in recent years. Many human services organizations use software programs that structure their documentation systems and use databases for information management that connect with local, state, and/or national systems. Human service professionals increasingly communicate with one another and with clients by e-mail and communicate less by telephone or face to face. Staff meetings and staff development programs also draw upon technology as PowerPoint® and other multimedia presentations are utilized in an effort to communicate more effectively. Most human service organizations have also developed websites, which they maintain to communicate with the community and to encourage the use of e-mail in business communications. This chapter focuses on the written and oral reporting required in human service agencies and the applications of computer technology to these important tasks.

Writing in Human Service Settings

Writing is an important component of virtually all human service organizations. Although you have become accustomed to the writing demands of being a student, the internship may be your first exposure to the writing demands of the profession. These demands can be particularly difficult because they often involve new styles and formats of writing and very limited time in which to produce a finished product. This difficulty may be compounded by a preference to spend the limited time you have available in your internship on tasks that you value more, such as time with clients or assigned projects. All of these factors conspire to put writing low on the priority list for many students as well as for many professionals. Like many students, human service professionals often consider the paperwork demands to be too time-consuming and sometimes become resentful of these responsibilities and resistant to carrying them out. As a result, lack of attention to record keeping has become a serious problem in many human service settings (Kagle, 1995).

Despite the many factors that might distract you from your writing responsibilities within your internship, do not allow yourself to develop bad habits now. Not only is writing an integral part of virtually any human service position, it is also a skill that you can develop and improve only with practice. Even if writing is not required by your agency, you would be wise to write mock reports and ask for your supervisor's feedback on them in order to develop your skills. Also, as you go about your work, take the time to read, analyze, and evaluate the written work of others: client records, agency reports, and other documents. Learn what you can by examining carefully those writing samples that strike you as particularly well written and incorporate what you learn into your own writing skills (Truitt, 1991).

◆ EXERCISE 9.1 ANALYSIS

In what ways does your internship organization use technology in its communications? In what ways does the technology make the work of the organization more efficient?

Less efficient? In what ways might technology be employed to further enhance communication within the organization?

The following case examples illustrate just how serious problems with documentation can become. Professionals who do not attend to the paperwork demands of their jobs can unintentionally harm the clients they are attempting to serve as well as their own careers.

CASE EXAMPLES: TWO CAUTIONARY TALES ON THE IMPORTANCE OF DOCUMENTATION

Harold had been employed as a human service professional for over 20 years in a sheltered workshop for adults with developmental disabilities. He was an effective worker and was well liked by his clients and colleagues as well by the agency's administrators. Over the years, he had become increasingly lax in maintaining his paperwork. His documentation of client contacts generally ran weeks, sometimes even months, behind. Harold told himself and others that "this paperwork stuff just isn't important. I'm too busy doing things that really matter to push a pencil for a bunch of bureaucrats."

A review of the agency by state authorities found many documentation deficiencies for which they officially cited the agency. As a result of this reprimand, certain components of the agency's funding were even jeopardized. Upon closer examination, agency administrators found that many of the documentation deficiencies were Harold's. When Harold's next performance evaluation was done, his supervisor rated his work unsatisfactory in this area and recommended that he receive no pay increases at all until he improved his work habits. In fact, Harold was even issued a warning that he could be terminated unless he changed his paperwork habits. He was given 90 days to show sufficient progress.

Harold felt angry and demoralized by his evaluation, yet there was little he could say in his own defense because his supervisor's evaluation was based upon objective data regarding his performance. Harold had deluded himself into believing that the paperwork requirements of his job really did not matter. Clearly, they mattered a great deal to the overall well-being of the agency. In some situations, client well-being can also be affected by poor documentation habits, as the following example illustrates.

Myra was a therapist in an adult mental health program that dealt with severely and persistently mentally ill clients. Myra had been working with a client, Carl, attempting to secure a placement for him in a long-term residential treatment program. She had made a formal referral to a particular program, which she documented in Carl's record but did not identify by name.

One Friday while Myra was away on vacation, Carl came into the program in a severe crisis. He needed an immediate residential placement outside his home. Because he was an unscheduled "walk-in," Sheila, the human service intern in the agency, was asked to work with Carl. Sheila's supervisor suggested that the residential treatment program to which he had already been referred might be the best bet for a speedy placement for Carl, especially if they should happen to have an emergency or respite bed. An additional advantage with this plan was that Carl could be spared a potentially upsetting transition from one program to another at some later time. Unfortunately, Carl was so disoriented that he could not remember the name of the program or anything about it.

Because the program was not named or clearly identified in the record, there was no way to locate it except through trial and error. Sheila spent much of the afternoon on the telephone calling various programs before identifying the one to which Carl had been referred. By this time, Carl had grown increasingly agitated and upset, but finally Sheila succeeded in making a placement for him. The process, however, took hours longer than it might have had the documentation in the chart been more detailed.

Harold's career was harmed by his lack of attention to his paperwork responsibilities. Carl's psychological state suffered (and probably Sheila's as well!) due to a careless omission in the record. Such events are all too common in human service settings, causing unnecessary hardship, inefficiency, and problems.

Written Reports Related to Direct Services to Clients

Writing about services to clients is probably the most frequent type of writing within most human service agencies. Documenting information about clients and the work that is conducted with them and on their behalf is an essential part of a human service worker's professional role. Knowing the purposes of documentation, general guidelines for doing such writing well, and some of the most common types of writing regarding services to clients will better equip you to meet the challenges of writing during your fieldwork.

Purposes of Documentation

If you see your writing responsibilities as just a useless bureaucratic requirement, you will naturally be reluctant to complete your written work. Far from being merely bureaucratic red tape, written documentation is an important component in the delivery of quality of services to clients and serves to protect your clients, your agency, and yourself. With this understanding, you will probably have greater motivation to do your written reporting and do it well. Following are some of the most important purposes of written reports and records regarding services to clients:

♦ *Written documents are necessary to provide continuity of services and care* (Kagle, 1996; Wilson, 1980). Clients sometimes need services when their usual worker is not available. A worker may be sick, at a professional meeting, or on vacation, or may have left the agency for another position. A clear, thorough record provides the information necessary for a different worker to provide services for the client efficiently and knowledgeably. In residential care facilities, the transition between workers occurs several times a day as workers report for various shifts. Smooth transitions rely upon workers leaving clear, thorough, and accurate information for one another. Another issue in continuity of care occurs when clients seek services in other agencies. With the client's permission, recorded information in such cases can be exchanged between the two agencies in order to reduce duplication of effort.

♦ *Written records help to ensure that workers have developed systematic plans and interventions for their clients' care* (Wilson, 1980). Writing a client care or treatment plan in which the client's history is summarized, current problems are identified, goals are set, and action plans are developed can be a time-consuming task. Nevertheless, developing such a plan in writing requires the worker to compile and organize relevant data, interpret it, and make decisions about how to proceed. Otherwise, overworked professionals might be tempted to rush from one client to the next without giving sufficient thought to planning services and interventions.

♦ *Written records provide legal documentation of worker activity* (Kagle, 1996; Wilson, 1980). Human service agencies and workers are held accountable for the services they provide. If a client should challenge the quality of care he or she received, the case record is considered legal evidence of professional activity. By the same token, action that is taken by the worker but not documented may be considered as not having occurred at all. Therefore, it is extremely important to document any tasks that you agreed to do on behalf of a client, all interventions you make on behalf of a client, and any tasks or assignments the client agreed to complete. It is also important to document outcomes of interventions, including progress made as well as less successful efforts.

♦ *Written records provide legal documentation of client needs* (Kagle, 1996). Just as it is important to document your own activity, it is important to document client behaviors, concerns, needs, and requests. Workers and agencies are sometimes put in the difficult position of defending why they did or did not take certain actions on behalf of a client. A record that shows that worker activity was based on observations of client behavior and/or concerns that the client reported offers a clear, credible rationale for the worker's decisions and actions.

◆ *Agency funding may be based upon the documented services delivered* (Kagle, 1996; Wilson, 1980). When agencies seek funding, whether it be from private grants or public revenues, they must submit statistics regarding the services delivered within the agency. Any service provided but not documented cannot be legitimately reported. Some funding sources do periodic site visits and conduct record reviews to verify that services that have been reported or billed were actually provided. The axiom, "If it isn't documented, it didn't happen," is applied in such reviews. As in the case example earlier in the chapter, agency funding can be significantly harmed by a worker's failure to record his or her work.

◆ *Written documents are used as a method to monitor the quality of the services delivered within an agency* (Kagle, 1996; Wilson, 1980). Agencies often use peer review committees to examine client records for information about the quality of the agency's work. Records might be reviewed to gather such information as whether services were delivered promptly, the number of worker contacts required to resolve client difficulties, the effectiveness of interventions being used by workers, and so on. Supervisors may also use records in similar ways to monitor the work of the staff members for whom they are responsible.

◆ **EXERCISE 9.2 ANALYSIS**

Interview a staff member in your organization about the paperwork demands of their role. How does the staff member feel about this aspect of the job?

For what purposes does the agency use written documentation? How are records used in the agency to enhance the quality of service delivery?

Does the staff member see ways in which the agency could make its paperwork processes more efficient?

Ground Rules for Documentation

Before you begin your work with documentation, it is helpful to understand some of the ground rules and guiding principles for writing about client care. Some of these ground rules are:

1. For client records to be useful they must be kept up-to-date. Write promptly to document any activity in the case so that another worker can be ready to assist the client in your absence, if necessary.

2. Be concise and to the point, but detailed enough to ensure that the necessary information is conveyed. Workers do not have time to read pages and pages of details. A good client record will present the relevant information in a clear, concise, organized fashion so that it can be read and understood fairly quickly. By the same token, the earlier case examples (see "Case Examples: Two Cautionary Tales on the Importance of Documentation") illustrated how important the inclusion of certain key details can be. Such critical details can be included without adding significantly to the length of the report.

3. Take care to document particularly clearly client needs and worker activities in an effort to meet those needs. These are central features in the work of any human service intervention.

4. Although most clients generally do not ask to see their records, clients legally do have access to any information regarding them in most settings. Be mindful of this whenever you write. Be honest and factual but tactful in your writing.

5. Do not use slang terminology in describing clients or their concerns. Although clients might use slang terminology in describing their own concerns, professionals should always use appropriate professional language. For example, in one report a student described a client with a drinking problem as a "lush." Although the client had used this term to describe herself, the student should have used more professional language.

6. Avoid subjective or judgmental language (Wilson, 1980). Use factual description instead. For example, rather than saying, "This client's house was disgustingly filthy," you might say instead, "There were eight large dogs inside the house. Dog feces and fleas were on the floor and furniture throughout the house."

7. Avoid diagnostic labeling (Wilson, 1980). Use factual descriptions instead. For example, instead of saying, "This client is anorexic," it would be better

to say, "This 21-year-old client is 5'6" tall, weighs 85 lbs., and may have an eating disorder."

8. Use person-first language and avoid referring to people as a diagnosis. For example, rather than reporting that the "client is a paranoid schizophrenic," it is preferable to say, "the client has paranoid schizophrenia."

9. Avoid or minimize sensationalizing language and phraseology such as "the client suffers from . . . ," "he is a victim of . . . ," or "she was stricken with . . . ." Use more straightforward, descriptive language such as the client "has" or "has been diagnosed with" a given illness or disorder.

10. Be clear in your writing about the sources of your information. Specify whether information is based on the client's report, your direct observation, or the report of a third party.

11. Clarify factual information as distinct from your impressions, opinions, and interpretations of the facts, although doing so might only involve a change of a few words. For example, it is one thing to say, "It is my impression that this client is not very motivated to find a job," and another to say, "This client is not motivated to find a job."

12. Always pay attention to the basic mechanics of good writing, including spelling, sentence structure, grammar, and organization. Once you have entered information into a client's record, it may be read by many professionals for years to come. Impressions are formed about you, your abilities, and your professionalism based upon the quality of your writing.

A Student's Reflections on Documentation

There was a major issue with a resident's family today. The doctor had changed a woman's medication during the course of the day. At the transition between second- and third-shift staffs, no one told the third-shift staff about the change. They gave the woman her old medicine that she had been taking. She then had a really bad reaction to the mix of medications. Understandably, her son was very upset about this. Although it was written in the record, the staff are supposed to write notes about any changes directly onto the log that stays on the counter in plain view in the nurses' station. It isn't efficient for the staff to sit down with a big stack of charts every shift to see what's been going on. This time the log system didn't work.

Common Types of Written Reports Regarding Direct Client Care

Each human service organization has its own unique methods of recording information about client care. Despite the differences from one organization to the next, in general there are a few common types of reporting that are present in most situations. Each of these types is briefly discussed here.

Client Care Plans (Treatment Plans)

Client care plans are reports that describe client needs and identify goals that the client and worker are striving to attain. The care plan, as the name suggests, lays out a plan for achieving the goals. Responsibilities of both the worker and the client are identified in the plan and tentative dates may be set for completion of various tasks or subgoals (Mehr, 1998). This report might be called a *treatment plan* in mental health facilities, medical facilities, or in other settings that follow a medical model.

The care plan is often one of the lengthiest documents in a client's record. Even so, it is important to remember that being concise and brief is a highly valued trait in all documentation. When a worker is reviewing a record to refresh her or his memory about a case or when a new worker picks up the record to become acquainted with a client before a first contact, neither has time to read pages and pages of detailed information. Finding the right balance between thoroughness and brevity takes practice and experience. By reading records within your agency, you can get some idea of what the norms are about the length of various documents within that particular setting.

Subheadings are generally used to organize the report and to help readers quickly identify any portion of the report they might be seeking. At minimum, the client care report should include the following information:

- *Identifying information* consists of such information as client name, gender, age, race, family status, employer/school, physical appearance, and socioeconomic status.
- *Presenting problem* records the immediate concern(s) that brought the client to the worker's attention.
- *History of problem* explains how the presenting problem(s) developed over time.
- *Social history* includes information regarding the client's development, family background, employment and/or educational history, legal history, medical history, and services previously received. Not all of these topics are relevant to all clients. Also, depending upon the agency's mission and goals, certain topics would get greater emphasis than others. For example, medical settings would pay more attention to medical history, whereas corrections facilities would place greater emphasis on legal history.
- *Goals* identify the outcomes the client and worker will pursue in their work together.
- *Plan* specifies the methods, activities, and strategies that will be employed to achieve the goals.

◆ EXERCISE 9.3 SYNTHESIS: LINKING KNOWLEDGE AND EXPERIENCE

Using the six headings above, write a care plan for a client in your internship setting. You might write about a client you have worked with or one you have observed. Alternatively, if your internship includes a field seminar, two students might perform a role play in class for the purposes of this exercise.

Identifying Information

Presenting Problem(s)

History of Presenting Problem(s)

Social History

Goals

Plan

Client Contact Notes (Progress Notes)

Contact notes are written after each encounter a worker has with a client, including office visits, home visits, school visits, telephone contacts, and so on. In residential care facilities, a note is written about each client at the end of each shift. Client contact notes tend to be fairly brief, perhaps a paragraph or two. They are sometimes referred to as *progress notes* or *summary notes*.

Although various formats are used, a particularly useful and common framework for organizing such a note is the SOAP approach. SOAP is an acronym that stands for the following components of a note:

◆ *Subjective (S)*. This portion of the note documents the client's report or that of the client's family or significant others—that is, the client's or other's subjective experience. For example, "Mrs. S. reported that her husband punched her and slapped her repeatedly last night and that she called the police this morning."

◆ *Objective (O)*. This portion of the note reports what the worker directly observed in the contact with the client. For example, "The left side of Mrs. S.'s face was swollen and badly bruised at the time of the interview. Using the phone in my office, she phoned the police to ask additional questions regarding how they would handle her report."

◆ *Assessment (A)*. The assessment gives the worker an opportunity to share any impressions, opinions, or interpretations relevant to the client contact. For example, "Mrs. S.'s face appeared as though it could have some broken bones. It is my impression that she may be in imminent danger if she returns home following her report to the police."

◆ *Plan (P)*. The worker reports any plans made with the client during the contact and/or interventions the worker plans to make on the client's behalf. The plan should include tasks to be completed by the client as well as by the worker. For example, "Mrs. S. planned to go directly to her doctor's office after our visit. I gave Mrs. S. information about Domestic Violence Services and encouraged her to call them as soon as possible, which she agreed to do. Secured Mrs. S.'s permission to contact the shelter to inquire about openings and admission procedures. Will call Mrs. S. at the home of her friend this afternoon to finalize plans for her immediate safety."

The use of subheadings is not generally required for contact notes unless they are lengthy. Whether or not the subheadings are used, the SOAP format provides an excellent structure to organize the worker's thoughts as well as provide an organizational structure for the note itself.

◆ **EXERCISE 9.4 SYNTHESIS: LINKING KNOWLEDGE AND EXPERIENCE**

Using the SOAP structure, write a summary note of a particular contact that you have had with a client or a contact that you have observed another worker conduct.

_____

Periodic Summaries

Periodic summaries, sometimes referred to as periodic reviews, are required at regular intervals within most settings. These reviews might be required at quarterly intervals (every three months), six-month intervals, and/or annually. These reports require workers to look at their work with a client over a longer period. Such reports are valuable because they require the worker to take stock of the progress that has or has not been made and to evaluate the effectiveness of the interventions being used. In the absence of such reviews, workers can get caught up in the day-to-day efforts being made with numerous clients and lose sight of the big picture.

Periodic reviews typically include the following:

- the time frame covered by the report,
- goals accomplished during this time,
- interventions the worker made during this time,
- new developments or problems that have arisen for the client during this time,
- goals that have yet to be accomplished, and
- plans for continued service, especially citing any revisions of the original care plan.

Periodic summaries are generally written in paragraph form without use of subheadings. Complex cases with many challenges and contacts might have fairly lengthy reviews of a page or more. Most reviews tend to be fairly brief, consisting of a few (1–3) paragraphs.

Closing Summaries

Closing summaries are written when services end for any reason. The closing summary should reflect an overview of the client's contact with the agency. It should include a statement of the presenting problem(s), a description of services provided, and a statement about the degree of progress made toward the various goals. The closing summary should also include information about the reasons why services are being terminated. In addition, any particularly significant events in the course of the agency's relationship with the client should be reported.

In writing this type of summary, keep in mind that if the client returns to the agency for services in the future, the closing summary should be helpful in reorienting the worker as to the client's status and to previous efforts to assist the client. This objective is especially important, of course, if a different worker assumes responsibility for the case upon the client's return. Also, the closing summary can become especially important if another agency requests information about the client following the termination of services.

The closing summary is the most logical source of current information in the client's chart in that it offers a comprehensive picture of the agency's work with the client and represents the most recent documentation of the client's situation (Kagle, 1996; Wilson, 1980). As is obvious from this description, the closing summary is fairly likely to be consulted as an important source of information in the future. Therefore, it is especially important that closing summaries be thorough, accurate, concise, and well written. You may have opportunities to write closing summaries as you terminate with clients toward the end of your internship.

◆ **EXERCISE 9.5 SYNTHESIS: LINKING KNOWLEDGE AND EXPERIENCE**

Write a periodic or closing summary of a helping relationship with a particular client over the course of your internship. If you have not had the opportunity to work directly with a client over time, you might read a client's record or write about a client you have observed. For the purposes of this practice assignment, you might summarize your work with a client to date although the duration of the relationship might only be a few weeks.

Most students find that the type of writing required in client records takes some practice. Continue to practice writing the various types of reports as much as possible. A good practice exercise is to observe another worker's contact with a client, write a summary, and then compare your summary with the one written by the staff member. Discuss with the worker any significant differences that you see between the two summaries. Such discussions will help to clarify any changes you might need to make in your methods of documentation. In most cases, documentation done by students must be cosigned by a supervisor (Simon, 1999). This safeguard helps to ensure that you are getting sufficient instruction about the particulars of how documentation is done within your setting.

Confidentiality is an issue that must always be considered when handling client information. No discussion of this topic can be complete without considering the ways that computer technology can potentially impact the confidentiality of client information. As discussed by Woodside and McClam (2006), three issues need to be addressed when computer technology is used in this way: the security of the data, the use of e-mail, and the security of the communication work site.

When agencies use databases that are connected to larger systems through the Internet, they must ensure that the data is secure through the use of such methods as encryption programs to scramble data en route and firewalls to prevent invasion of the data by outside users. In use of e-mail, professionals should be mindful that e-mail is not a secure system of communication. Messages can sometimes be sent to the wrong address inadvertently, and other people besides the client may have access to the e-mail account. Nevertheless, many clients want the convenience of communicating with their human service workers by e-mail. Whether to assume this level of risk should be the client's decision. Human service professionals should secure signed consents from their clients, specifying the risks involved and granting them permission to communicate with the client by e-mail. The client should also make the choice as to whether it is acceptable for professionals to communicate with one another about them by e-mail and give written consent if they so choose.

Finally, technology has changed work practices significantly in that professionals can now work from home or other remote sites far more easily than in the past. Each time a professional chooses to work outside the office, special care must be taken to secure documents and notes used in the process. Home computers are generally shared with family members and even visitors to the home. Human services professionals who work from home must ensure that family members and visitors do not have access to electronic files regarding clients or the handwritten notes that might have been used in creating these files (Woodside & McClam, 2006). Similarly, e-mail communication that is carried out from home or other sites must also be secured. Technology has streamlined many tasks for human service professionals and has great potential to facilitate communication with clients and colleagues. With these conveniences, however, come special challenges and responsibilities to safeguard client confidentiality in this electronic age.

 ◆ **EXERCISE 9.6 ANALYSIS**

To what extent does your organization incorporate computer technology in its documentation related to client services? What particular safeguards, if any, are taken to ensure client confidentiality within these systems?

To what extent do clients have control over the way such information is handled?

Discuss the advantages and disadvantages of your agency's use of computer technology.

Written Reports Related to Administrative Services

Students and professionals in administrative roles may write as much or more than do those in direct service roles, but their writing tends to be of a different type. If you are in an administrative internship placement, the writing requirements that you encounter might be quite varied. Administrators often write for the purpose of internal communication within the agency as well as for communication with external audiences, such as other agencies, funding sources, community organizations, media outlets, and the community at large. A few of the most common types of reports are discussed in the following sections.

A Student's Reflections on Administrative Reporting

As one of my projects, I was asked to take charge of a major statistical report that they have to do every year. It involves retrieving data from various computer files, organizing it into tables and charts, and writing a narrative report that describes and interprets the information. The report is important to the agency's accreditation because it documents staff training in every department and division. I am pleased to say that I have been doing pretty well on this project and that I have gotten lots of praise and pats on the back for my work. Who says that paperwork is a thankless task?

Reports About Agency Programs

Reports about the agency and its various programs may be prepared by program directors or coordinators and submitted to agency directors and other administrators for the purpose of intra-agency communication. Such reports might include statistical information regarding the clients served as well as descriptive information about the programs, services, and administrative processes within the unit. These reports often identify strengths and weaknesses in the program or unit and make recommendations for improvements and changes. Program evaluations, in which administrators assess the extent to which program goals have been accomplished and the quality of the agency's functioning in the course of meeting these goals, are routine in most organizations. Such reports are usually written annually and form the basis for program planning and implementation for the next year (Weiner, 1990).

Time Interval Reports

Time interval or periodic reports are required within many agencies. These reports are sometimes referred to as *process evaluations* (Lewis, Lewis, & Souflee, 1991) or *interim reports* (Seidl, 1995) because they provide ongoing information about the agency's activities that is used to monitor how effectively the various units within the organization are achieving the agency's goals month-by-month or quarter-by-quarter. Each unit within the organization might be required to submit such a report to the agency director monthly or quarterly in order to ensure that each unit's work is staying on track. These reports may be fairly comprehensive, covering such topics as number and types of clients served, staff–client ratios, specific services rendered, financial expenditures and acquisitions, special program initiatives, and so on for the period covered.

Reports to the Board of Directors

Reports to the board of directors routinely include financial information and progress reports from various units and/or the organization as a whole. Because the board of directors is responsible for the agency's performance, it must monitor activities closely (Gelman, 1995). Written reports are often used for this purpose. At times, additional reports are generated exclusively for the purpose of communicating with the board. Examples of such reports might include formal proposals for new programs, for program modifications, or for policy changes. Such reports are submitted to the board of directors as a first step in seeking board input and approval. Reports of various kinds (both written and oral) are made to the board at most regular board meetings. The by-laws of agencies dictate such meetings at set intervals, such as monthly, every two months, or quarterly.

◆ EXERCISE 9.7 ANALYSIS

Do some research in your agency to identify the various types of writing that administrators are required to do.

Which of the types of writing discussed here are generated within your setting and at what intervals?

What additional types of reports and writing are required?

Compile samples of the various types of administrative reports and writing required within your agency and critique their usefulness. Does the administrative writing and reporting that is done contribute to the agency's effectiveness? If so, how? If not, discuss what changes might be made to use these efforts more effectively.

Funding Requests

Funding requests are written in agencies of all types. Some agencies may seek funds from tax revenues through filing proposals to various elected boards and/or government representatives. Other agencies may seek funds by making requests to such sources as United Way, private donors, or grant sources, both public and private. Private nonprofit programs might even submit funding requests to corporations. Two common types of documents for fundraising purposes are case statements and grant requests.

The case statement is important in fund-raising from private sources. It serves to cultivate interest and support from potential donors. This document typically includes a description of the organization seeking funds, a convincing argument as to why support is warranted and for what purpose, and an explanation of how contributions will be used to address the needs described. This statement is then used in a number of ways. It may be used as an outline by volunteers as they speak with potential donors, as an initial mailing to prospective donors prior to a person-to-person visit from a campaign volunteer, or as a handout to audiences following fund-raising presentations (Turner, 1995). The case statement is an important fund-raising tool in those organizations that engage in soliciting donations directly from donors.

Over the last two decades, grant writing has become increasingly common in human service agencies. Although the various sources of grants might require slightly different types of proposals, grant requests generally include the following information: a description of the need or problem that the funding will be used to address, evidence to support the existence of the need or problem, a description of the project for which funding is being sought, the specific goals and objectives of the program, the strategies or plan of action that will be used to achieve the project's goals, the methods that will be used to evaluate the program, and a budget or estimation of the program's cost (Lewis, Lewis, & Souflee, 1991).

Inquire about how you can become involved in any fund-raising and grant-writing efforts in the agency so that you can learn as much about the process as possible. Human service students can considerably enhance their skills and employability by developing expertise in these areas.

A Student's Reflections on Grant Writing

I am working on a grant for a new position in the agency. I have researched and located potential funding sources and am currently engaged in the process of writing the grant itself. The agency needs to set up a one-on-one mentoring program in the community to work on drop-out prevention, but no one has the time or money to get one going, so the grant I am working on will try to secure funding for this. It's exciting to be working on a real grant, especially since the grant would create a job I would love to have if it is funded. I have an extra incentive thinking that I may be writing up my own first job opportunity. Whether or not this works out, I now know that I can write a real grant. I think employers will be interested in that.

◆ EXERCISE 9.8 ANALYSIS

What efforts, if any, does your agency make to raise funds from the community, foundations, or state and federal agencies?

What written materials do they produce toward this end?

Locate recent examples of case statements and/or grants used in your agency's fund-raising recently and critique their effectiveness below.

Public Relations

Public relations has become a critical component of administrators' roles in human services organizations. Written materials comprise one of the most essential cornerstones of the public relations efforts within many agencies. Such activities as writing press releases, creating brochures, and producing agency newsletters are now commonplace within human service agencies of all types and sizes. These efforts are important in maintaining community awareness of programs, marketing the services that the agency offers, and managing the public image of the organization (Weiner, 1990). Many human service organizations cultivate relationships with local journalists and newscasters to facilitate their ability to promote the organization locally.

Agency websites in recent years have become another important public relations tool for human service organizations. These websites are used to educate and inform the community about the agency's mission and services, communicate with clients, recruit volunteers, solicit donations, post position vacancies, and promote agency fund-raisers and programs. Although websites provide a wonderful communication tool within the community, it is important to recognize that many human service clients do not have access to technology or even to community newspapers. A number of populations are likely to be affected by the digital divide, including minorities, the elderly, and people in rural areas. The term *digital divide* refers to the fact that accessibility to computer technology varies according to socioeconomic status. In general, it is the poor who are most negatively affected by the digital divide (Fridman, 2000). Organizations must be committed and creative to design public relations efforts that will reach into all segments of the community.

 ◆ **EXERCISE 9.9 ANALYSIS**

Review the agency's website and review written materials that are produced to enhance its public relations efforts. What are the strengths of these resources? What are their weaknesses?

If your agency does not have such resources, find out what you can about why this is the case. What are your ideas about how the agency might benefit from the creation of such resources?

What methods might your organization use to reach populations who may not have access to technology, community newspapers, or other commonly used methods of public relations?

If your agency is one that does not currently produce any public relations materials, you might offer to produce some samples as a special project during your internship. You might try your hand at creating a draft for an agency brochure, for example, and share your efforts with your supervisor and other staff members. If your agency produces such materials, inquire about how you might get involved in these efforts. For example, you might write an article for the next edition of the agency's newsletter. Students are often able to provide positive, fresh, and interesting perspectives on the agency in these articles. Such efforts on your part will give you an excellent opportunity to practice and develop new skills while providing a wonderful service for your agency.

Some agencies do not have websites because they lack the resources, time, or skills to do so. If this is the case with your internship organization, you might explore whether it would be worthwhile to connect the organization with the computer science program at your college or university. Many computer science programs have web design courses. Developing agency websites can provide excellent service-learning opportunities for students in these courses. As a human services intern, you might partner with the computer science students to provide more in-depth knowledge of the organization and its clients.

Oral Reporting

In addition to written reports, oral reports are commonly required of students and professionals in human service settings. Professionals in administrative roles frequently present information to staff groups, community groups, potential funding sources, and boards of directors. Professionals in direct service roles speak before their peers and/or supervisors to present case material, receive consultation, and conduct education and training for one another. Such presentations are often done in formal, regularly scheduled meetings sometimes referred to as *staffings, case conferences, review team,* or *grand rounds.*

These presentations are a valuable part of any organization's functioning and serve a number of purposes. They ensure that workers are receiving adequate oversight and the input of other professionals as they conceptualize and intervene in complex human lives. Additionally, such presentations can constitute a form of continuing education for staff as they learn from hearing their peers present on various issues in their fields of expertise.

In planning and preparing for oral reporting and speaking, it is important that you consider the audience for your remarks, bearing in mind its characteristics. Although it is important to not make too many assumptions about your audience, it is necessary to think through who it will be. Considering issues such as age, ethnicity, background, gender, knowledge, and relevant experiences helps you develop your ideas in keeping with its interests and needs (Tollefson/Peterson, 2004). For example, reporting to your supervisor individually may be different than reporting to all of the professionals in your department at a staff meeting or case conference. Similarly, speaking before a client group may require special sensitivity as opposed to speaking before a group of professionals. Speaking to a relatively unknown community group may require more formality than speaking before people you know. Group size is also an important element of audience analysis. Presenting before a large group will require a different style and set of skills than does speaking before a smaller group.

As you develop your oral presentation, clearly identify your goals. What do you want your audience to know at the end of your presentation? What do you want it to do with the information (Tollefson/Peterson, 2004)? In internal staff groups, you are often presenting information simply to get the facts of the situation before the group for the purposes of consultation and decision making. In such cases, you would want to paint a clear, accurate, comprehensive, and descriptive picture of the situation so that decision making can be based on good information. In making community presentations, your goals may be educating the community about a given issue, raising funds for an organization, or mobilizing a group for action. Again, the goal you are trying to achieve should shape what you say and how you say it.

You must also consider the type of environment in which you will be making your presentation. Will it be a large room or a small room? Will you be in a formal, lecture-style room with a lectern or podium? Or will you be seated in a circle with the audience members? What barriers might there be that you will need to overcome? How might you arrange the room to best advantage for your presentation? The comfort of the room, external noise and interruptions, and even the time of day when you present can impact your audience's attentiveness. The more you can know in advance about your audience and your environment, the more effectively you can plan and the better prepared you can be.

◆ EXERCISE 9.10 PERSONAL REFLECTION: OBSERVATION OF SELF AND OTHERS

Under what circumstances have you presented information orally to other professionals, community groups, staff members, and others during your fieldwork?

How comfortable were you in these situations?

Most students and many professionals need to practice their oral presentation skills as much as possible. Can you identify additional opportunities that you might pursue in order to practice your oral presentation skills during your internship?

A Student's Reflections on Oral Reporting

There was a meeting of the Board of Directors last night, and my supervisor asked me to come for a few minutes and tell them about the work I've been doing. I don't like talking in front of groups, but I went. I took an outline of the different things I have been doing and what I wanted to say about each one. This may seem silly, but I had even practiced. As I drove along in the car over the past few days, I would talk out loud, pretending I was doing my talk. I think things went well. I had the most trouble at the end when they asked me questions. I hadn't really prepared for this, and I heard myself saying "uh" a thousand times as I tried to get my thoughts together. I can do pretty well when I'm rehearsed, but I need to get more practice talking off the cuff.

Students frequently report that presenting information orally is an anxiety-provoking experience for them, and understandably so. Oral presentations can be particularly challenging because both verbal and nonverbal aspects of the communication must be handled effectively. To enhance your presentation skills, it may be helpful to review some basic principles about presenting information effectively.

To understand the skills involved in oral presentation, it is important to remember that face-to-face interactions carry information in three categories: vocal expression, body language, and verbal expression. One researcher has suggested that as much as 93 percent of the total message is conveyed through the nonverbal avenues of vocal expressions and body language (Mehrabian, 1981). Vocal expression includes such factors as rate of speech, voice pitch, volume, and tone of voice. To evaluate your effectiveness in this area, you might arrange to record yourself in a meeting or presentation. Either audiotape or videotape is useful in gathering information about your vocal expression. In general, you should strive to speak in a strong voice, loud enough to be heard by everyone in the room without their having to strain to hear you. You should speak slowly enough to be understood but not so slowly that others become bored or impatient.

Anxiety can sometimes tighten the vocal chords, causing the speaker's voice pitch to become higher or even creating a quiver in the vocal chords. Anxiety can also influence the speaker to rush through the information, speaking far too quickly to be easily understood. If you feel that anxiety interferes with the quality of your presentations, it may be useful to practice some relaxation methods such as taking slow, deep, even breaths, inhaling slowly through the nose, and exhaling slowly through the mouth. This breathing pattern helps to induce a relaxed response and can be practiced fairly inconspicuously anywhere.

Nonverbal communication includes not only vocal expression but also body language and facial expressions as well. Although a great deal of attention is given in human service education to nonverbal communication as it relates to interactions with clients, you might have given less thought to its relevance in other areas of professional communication.

Many of the same principles and guidelines that apply to making effective presentations apply to interacting with clients effectively. For example, one of the most significant factors in making effective presentations is facing the person (or people) whom you are addressing squarely and making eye contact. As you know, this same communication skill is also required in effective interaction with clients. A student may maintain eye contact perfectly in a helping interview, but it is not unusual to see this same student stare directly down at a pad placed in his lap or twirl her hair anxiously in front of her eyes when presenting information before a group or in supervision. In addition to maintaining eye contact, avoid fidgeting and other anxious habits when presenting. Common examples of such behaviors include shaking the foot, swinging the foreleg, rocking or swaying from side to side, playing with a pen or paper clip, doodling on a pad, twisting a ring or earring, and handling the hair. All of these behaviors convey anxiety and distract listeners from hearing what you have to say. Focus on standing or sitting straight, speaking up, and maintaining eye contact.

Finally, verbal expression is obviously important because it carries the intended content of the message. To present a clear verbal message, you will need to plan and organize what you want to say ahead of time. Whether you are presenting a case in staffing or making a presentation before the agency's board of directors, you need to think ahead of time about the major points you wish to make and prepare notes that can act as prompts to keep you focused and organized as well as to jog your memory, if necessary (Halley, Kopp, & Austin, 1998). Under no circumstances should you simply read your remarks from a prepared draft. Although it may be reassuring to write word for word what you want to say, this might then tempt you to just read what you have prepared. It is generally more helpful to speak from notes, even if they are quite detailed.

Although a speaker might be very knowledgeable and well prepared for a presentation, the effectiveness of any presentation can be quickly undermined by counterproductive language habits. For example, the credibility of a presentation is weakened if the speaker repeatedly uses qualifiers such as "I think," "sort of," and "I guess." Similarly, speakers whose sentences are sprinkled with such phrases as "uh," "well," "um," "like," and "y'know" are not only difficult to listen to but also come across as lacking in confidence and poise (Weaver, 1993). As suggested earlier, audiotaping or videotaping is perhaps the best way of identifying such language patterns in yourself (Truitt, 1991).

Other guidelines regarding the verbal component of presentations closely parallel those discussed earlier for written communication. Just as in your writing, use correct grammar and avoid the use of slang language. Also avoid using subjective or judgmental language, focusing instead upon factual data to make important points. Acknowledge when you are making a generalization or sharing a personal impression and back up your statement with specific facts and examples. Although you might have strong feelings about a given client or situation, your oral presentation will be more effective if you avoid highly emotional displays, adhering instead to factual, descriptive information. Similarly, a speaker presents as more accurate and credible by avoiding extremes of verbal expression. For example, describing a situation in terms such as "always," "never," "all," or "none" should be avoided unless it is *literally* true (Weaver, 1993). Figurative use of such terms can reduce the speaker's effectiveness by creating the impression that he or she is exaggerating or holds rigid or extreme views. Of even more concern is the possibility that a situation or person may be inaccurately and unfairly described through the use of such terminology.

Because oral presentation skills are clearly quite important, it is a good idea to try to get some objective feedback about your own oral presentation skills during the course of your fieldwork. You might arrange to audiotape or videotape yourself when presenting, then review the tapes to evaluate your skills. It is also helpful to ask for feedback and suggestions from your supervisor and other co-workers who have observed you when presenting information within the agency.

 ◆ **EXERCISE 9.11 ANALYSIS**

Attend a meeting of the board of directors of your organization or another type of administrative meeting, if possible. These meetings often include a combination of oral and written reports. Observe the various participants in the meeting and the documents that are distributed.

Describe here the particular strengths and weaknesses you observed in oral and written communication in this meeting.

Conclusion

Many students are surprised to learn how much time and effort human service professionals put into writing and reporting. Although students are generally well prepared to provide client services, they may be less prepared for the reporting functions of human service worker roles. Skills in verbal communication, nonverbal communication, and writing are required to handle these requirements effectively. These skills can be developed through ongoing effort, practice, motivation, and self-evaluation.

FOR YOUR E-PORTFOLIO

Submit some samples of your professional writing and reporting to your e-portfolio. Include a range of different kinds of writing, such as writing related to client care (e.g., client care plans, summary notes, periodic summaries), and writing for administrative purposes (e.g., grants, public relations materials, administrative reports). Also submit information regarding at least one oral report that you have done in your internship. Your submission might include an outline or summary of your presentation along1 with an analysis of your audience and the goals you sought to achieve through this presentation. Additionally you might include a digital video clip of yourself giving an oral presentation.

After reviewing these submissions to your portfolio, comment on their effectiveness. What do you see as your strengths and weaknesses in written and oral reporting? Conclude this portfolio entry by stating your specific goals for improvement and your strategies for accomplishing those goals.

References

Fridman, S. (2000, June 19). Money, not race, underlies digital divide. *Newsbytes*. Retreived November 20, 2005 from http://www.computeruser.com/news/00/06/19/news18.html.

Gelman, S. (1995). Boards of directors. In R. Edwards & J. Hopps (Eds.), *Encyclopedia of social work* (pp. 305–312). Washington, DC: NASW Press.

Halley, A., Kopp, J., & Austin, M. (1998). *Delivering human services: A learning approach.* New York: Longman.

Kagle, J. (1995). Recording. In R. Edwards & J. Hopps (Eds.), *Encyclopedia of social work* (pp. 2027–2032). Washington, DC: NASW Press.

Kagle, J. (1996). *Social work records* (2nd ed.). Prospect Heights, IL: Waveland Press.

Lewis, E., Lewis, F., & Souflee, F. (1991). *Management of human services programs* (2nd ed.). Pacific Grove, CA: Brooks/Cole.

Mehr, J. (1998). *Human services: Concepts and intervention strategies* (5th ed.). Boston, MA: Allyn & Bacon.

Mehrabian, A. (1981). Silent messages: Implicit communication of emotions and attitudes (3rd ed.). Belmont, CA: Wadsworth.

Seidl, F. (1995). Program evaluation. In R. Edwards & J. Hopps (Eds.), *Encyclopedia of social work* (pp. 1927–1932). Washington, DC: NASW Press.

Simon, E. (1999). Field practicum: Standards, criteria, supervision, and evaluation. In H. Harris & D. Maloney (Eds.), *Human services: Contemporary issues and trends* (pp. 79–96). Boston: Allyn & Bacon.

Tollefson/Peterson. (2004). How to make your speaking easier and more effective. University of California Berkley, Division of Undergraduate Education. Retrieved November 13, 2005 from http://teaching.berkeley.edu/speaking.html.

Truitt, M. (1991). The supervisor's handbook: Techniques for getting results through others. Shawnee Mission, KS: National Press Publications.

Turner, J. (1995). Fundraising and philanthropy. In R. Edwards & J. Hopps (Eds.), *Encyclopedia of social work* (pp. 1038–1044). Washington, DC: NASW Press.

Weaver, R. (1993). *Understanding interpersonal communication* (6th ed.). New York: HarperCollins.

Weiner, M. (1990). *Human services management: Analysis and applications.* Belmont, CA: Wadsworth.

Wilson, S. (1980). *Recording: Guidelines for social workers.* New York: Free Press.

Woodside, M., & McClam, T. (2006). *Generalist case management: A method of human service delivery.* Belmont, CA: Thomson Brooks/Cole.

Chapter 10

Managing Your Feelings and Your Stress

A human services internship can feel like an emotional roller coaster. The excitement and stress of being in a new and unknown situation with numerous responsibilities can make an internship in any discipline emotionally challenging, but compounding this situation is the intensely personal and interpersonal nature of most human services internships. Working daily with difficult human problems prompts emotional reactions in even the most experienced professionals. Within the span of just a few minutes, you might feel both the gratification of knowing that you have helped a client and the frustration and helplessness of feeling that you have little or nothing to offer. You might experience genuine warmth and empathy for one client followed by irritation and frustration toward the next.

All these experiences are within the scope of a normal day's work in human services. The challenge for you as an intern, and throughout your career, is to manage and deal with these emotions in a manner that is both professional and healthy. This chapter focuses upon specific skills and strategies to help you to manage your feelings and your stress effectively.

A Student's Reflections on Managing Feelings and Stress

I am busy every single minute. This agency is so understaffed it is pathetic. I took on the responsibility because I'm here to learn, and I thought it would be good for me. Now I think I have way overextended myself. I blew up the other day at Sam,

continued

---

one of the guys who works here. I am in charge of planning a big field trip for all of the kids, but I thought it was going to be a team effort. Sam didn't do the things he agreed to do, which then put me in a crisis. I don't have room for three or four more things on my list of things to do. Not to mention that I have been really worried about one of the girls who has been coming to the program. She seems really down, and I think I have smelled beer on her breath a few times lately. She is only 13 years old and a sweet kid. I know there are a lot of alcohol and drug problems in her family. I want to approach her about it but the time has to be right. The last thing she needs is for me to scare her away from the program. We are her main support right now. I am completely stressed-out.

Developing Self-Awareness

Perhaps the most fundamental skill in managing your emotions is self-awareness. Self-awareness is the best place to start in managing your emotions because it enables you to recognize your feelings as they occur. Noticing and honestly acknowledging your feelings, reactions, thoughts, and behaviors is the first step in confronting and dealing with your emotions in a responsible, professional manner. Although the focus of this chapter is on managing feelings, you will find that monitoring your thoughts and behaviors can be helpful in recognizing your emotions because they are often important sources of information about your feelings. Thoughts and behaviors can sometimes act as signposts, leading you to recognize emotions that may be present but less obvious, accompanying or underlying your thoughts and behaviors.

Developing and maintaining self-awareness during your internship requires you to cultivate a nondefensive, open psychological posture, which can be difficult when you are in the midst of anxiety-provoking or other emotionally intense experiences. Throughout the day, as you find yourself becoming tense or anxious, practice simple, unobtrusive relaxation methods such as deep breathing and relaxing your shoulder muscles. Practice self-observation throughout the day, trying to stay in touch with your emotional reactions to various events. Do not feel that you should necessarily act on these emotions or even express them. In fact, an important goal of self-awareness is to gain greater control over your feelings so that you will not act on them in a manner that is unprofessional or counterproductive. Work on simply observing your feelings, thoughts, and behaviors in various situations as a method of learning more about yourself and learning to monitor your emotions.

◆ EXERCISE 10.1 PERSONAL REFLECTION: OBSERVATION OF SELF AND OTHERS

What are the prevalent emotions that you have observed in yourself during the course of your fieldwork? What are some of the most difficult emotions for you to deal with productively? What specific events or circumstances triggered these emotions in you? What strategies have you used to deal with these emotions productively?

Developing Self-Understanding

Building upon self-awareness, you can work toward enhancing your self-understanding as well. Self-understanding goes beyond simply identifying your feelings and involves recognizing the sources of those feelings. Let's imagine that you have just participated in a meeting in your field agency that you found to be very uncomfortable. By the end of the meeting, you were feeling quite distressed, anxious, and somewhat annoyed. Although this self-awareness is an essential first step in managing your emotions, important questions remain to be answered. Where do these feelings come from? What particular events in the meeting might have triggered these emotions? Were other participants in the meeting inappropriate in their behavior or are your feelings the result of a more personal sore spot for you? Do your feelings stem directly from events in the meeting or are other issues in your life a factor? Do the feelings suggest that there is an issue or problem that needs to be resolved? If so, is it an issue within yourself? A problem within your internship? An interpersonal conflict? A combination of all of the above?

While self-awareness relies mostly on self-observation, self-understanding relies on deeper reflection. It may take some careful thought and analysis to identify the issues involved in a particular disturbing event. Private time for reflection, introspection, and just plain thinking is essential in order to develop this type of self-understanding. Writing in a journal or processing significant events using the Integrative Processing Model (discussed in Chapter 4) are ideal tools for developing greater self-understanding. Also, talking with your supervisor or another good active listener can help you to develop greater insight into your emotional reactions.

◆ EXERCISE 10.2 PERSONAL REFLECTION: OBSERVATION OF SELF AND OTHERS

Identify and discuss an instance during your internship in which you were able to move from an awareness of your feelings (self-awareness) to understanding the source of your feelings (self-understanding). What methods have you used to cultivate self-awareness and self-understanding during your internship?

Developing Assertiveness

Once you have understood the source of your feelings, assertiveness can be especially useful at times in managing your emotions effectively. In some cases, examining and understanding your emotions helps you to identify problems that need to be solved or issues that need to be addressed in your internship. In these cases, the next step must involve expressing those needs to others so that problem solving can occur.

Assertiveness is simply the ability to straightforwardly communicate your thoughts and feelings, respecting your own needs as well as the needs of others (Alberti & Emmons, 1995). As such, assertiveness is an essential tool for human services workers to possess as they engage in dealing with emotions and solving problems. Assertiveness involves the use of clear and confident "I-statements" in expressing what you want, need, or feel. Some people have difficulty allowing themselves to be assertive because they misunderstand what assertiveness is, believing that assertiveness requires them to be selfish, demanding, or rude. These behaviors are not assertive behaviors but aggressive behaviors, which are discussed later in this chapter. Remember that assertive communication recognizes and respects the rights of others.

If you find it difficult to communicate assertively, it might be helpful to practice the empathic-assertive model of communication. Within this model, the speaker first expresses the other person's perspective before expressing his or her own perspective (Lange & Jakubowski, 1976). For example, in speaking with your field supervisor, you might say, "I realize that you are concerned that this project might be too much for me, but I feel very confident that I can do it well. I would very much like to take it on." Note that this statement reflects the feelings of both parties in the conversation. Nonetheless, the speaker makes a clear, straightforward request.

Assertive communication includes some important nonverbal components in addition to the verbal components previously discussed. Even the best assertive statement can be rendered ineffective if accompanied by incongruent nonverbal messages. Assertive nonverbal behavior includes maintaining eye contact, standing or sitting with an erect posture, facing directly the person whom you are addressing, and speaking in a firm, clear voice.

Observing good models of assertive communication has proved to be an effective tool in helping less assertive people become more assertive. Simply watching another person handle a situation assertively can be a powerful educational experience, potentially expanding your own behavioral repertoire. Take a moment and try to identify a good role model for assertive communication, perhaps someone whom you have observed communicating assertively in your fieldwork setting. Imagine yourself behaving in a similar manner.

◆ EXERCISE 10.3 PERSONAL REFLECTION: OBSERVATION OF SELF AND OTHERS

Discuss at least one situation during your internship that you handled assertively. How effective were you in maintaining an assertive stance? How might your assertive response have been improved? What lessons can you learn about assertive communication from the model that you selected in the preceding paragraph?

In contrast to assertiveness, both nonassertive behavior and aggressive behavior are ineffective means of dealing with your emotions and do not lead to problem solving. Nonassertive or passive people tend not to speak up once they identify important needs and feelings. Instead, they are likely to keep their feelings to themselves or perhaps even hide their feelings from themselves. As a result, the problematic situation continues and possibly even grows worse. In addition to the original problem, the individual tends to develop an extra burden of anger and resentment as the troubling situation continues. Consequently, over time, the feelings of the nonassertive person can grow even more intense and difficult to manage. The nonassertive person who does speak up about a concern may be more likely to talk to a third party rather than speak directly with the person who needs to hear the concern. This, too, tends to create more problems rather than solve them, as the following example illustrates.

CASE EXAMPLE: A NONASSERTIVE RESPONSE TO A PROBLEM

During her internship, Latoya had been uncomfortable with a particular co-worker, Shirley. Shirley had virtually ignored Latoya throughout her internship. She never once asked Latoya to work with her or even to observe her work. Because Shirley's area of work was of particular interest to Latoya and related to her career goals, this was of real concern to her. After a few weeks, Latoya began to express her frustration and irritation about Shirley's cold shoulder to co-workers and to other students in the agency. Within a few days, Shirley heard through the grapevine that Latoya had been talking about her.

Shirley (who had no difficulty being assertive) confronted Latoya in an appropriate manner, expressing her concern that Latoya had apparently been speaking of her in a derogatory manner to others in the agency. Latoya quickly apologized to Shirley and left the room, still not having the courage to talk with her directly about the situation. For the remainder of the internship, Latoya was even more uncomfortable in Shirley's presence and tried to avoid her as much as possible. Although Latoya otherwise had a very beneficial internship experience, she left at the end of the term with great regrets about how she handled the situation with Shirley. Worst of all, she never had an opportunity to learn about Shirley's job. In this situation, as in many, nonassertive behavior did not solve the problem and, in fact, made matters even more tense and uncomfortable.

Aggressive communication, though quite different from nonassertive communication, tends also to make situations worse rather than solve problems. Aggressive behavior involves expressing feelings and needs in a manner that attacks others or is disrespectful of others' needs. This pattern may result in the escalation of conflict rather than in problem solving. Also, the person who employs aggressive communication risks alienating others and damaging relationships (Alberti & Emmons, 1995).

Let us imagine that in response to Shirley's assertive statement of anger described earlier, Latoya had gone on the attack. Feeling wronged, Latoya might have responded by saying, "How dare you accuse me of causing a problem between us! You have done nothing but be hostile and unhelpful toward me since the minute I walked in the door!" Notice that this statement is accusatory and attacking. No I-statements are used. Instead, Shirley is angrily blamed for the situation. This is not an assertive response, but an aggressive response. From this example, it is probably obvious how aggressive responses can escalate problems and block potential avenues of problem solving.

Developing Conflict Resolution Skills

Although most people do not enjoy conflict, it is an inevitable part of human relationships. All organizations, including human service organizations, experience conflict (Halley, Kopp, & Austin, 1998). In human service agencies, conflicts can and do occur between workers in the same agency, between workers from different agencies, and even between workers and clients. When conflicts occur, they can become major stressors until they are resolved.

If you should experience conflicts during your internship, try to view them as opportunities to develop and practice conflict resolution skills. Do not let the conflicts fester for long, because they tend to weigh you down emotionally and distract you from other important areas of potential learning in your internship. Learning to resolve interpersonal conflicts can be an important part of reducing your stress and managing your emotions effectively, both in your internship and later in your career. Developing and practicing these skills is particularly worthwhile because such skills are highly valued in employees in virtually any work setting. Moreover, conflict resolution skills are especially valuable to human services professionals because they form the basis of many helping strategies and interventions (Hagen, 1999).

A Student's Reflections on Conflict Resolution

In my family, conflict is something to be avoided. It's not OK to be openly angry or even to disagree. For many years, I thought that I must have a perfect family because we never had any conflict. Because of my background, now I have a great respect for people who are able to handle conflict well. One thing that has impressed me about this agency is how upfront everyone is with their feelings. If they disagree with one another, they say so. Usually it is a friendly disagreement, but at times it can even get emotional. The staff seems committed to talking things through and working things out. Although I haven't personally been involved in a conflict situation here, I am learning a lot from just watching these people. They practice what this profession preaches about the importance of communication, respect, and finding win-win solutions.

Your ability to resolve conflicts will rely in part upon your mastery of skills discussed earlier in this chapter. Being aware of your feelings, understanding yourself, and communicating assertively are foundational skills in resolving conflict. In conflict resolution, you use these skills to state what you want and how you feel. It is equally important that the other person(s) do the same and that you listen respectfully to their needs, wants, and feelings. Negotiations should focus on efforts to find win-win solutions to the extent possible.

Win-win solutions are those that satisfy the needs of all parties. They require the parties involved to engage in a problem-solving process in which the needs and interests of all parties are explored and possible solutions that might satisfy the interests of all parties are brainstormed and considered. Win-lose negotiations occur when the parties doggedly pursue an outcome more favorable to themselves than to the other people, employing such strategies as tests of will, manipulation, or power plays. Although win-win conflict resolution takes time and patience, it is most beneficial to preserving goodwill in long-term relationships (Johnson, 2006). Therefore, this method is advisable for co-workers in human service agencies who are likely to work together for some time. Although you may be in the organization a relatively brief time, the relationships you form there can have lasting effects on your career. It is wise to treat them as long-term relationships.

Effective conflict resolution requires effective anger management. Letting anger take control can easily destroy the possibility of constructive problem solving. Managing your anger does not mean that you do not feel it or express it. Rather it means that you contain it so that your efforts toward problem solving can stay on task (Johnson, 2006). If you feel the need to express your anger, do so constructively, using assertive rather than aggressive communication (see "Developing Assertiveness" earlier in this chapter). Also, when you feel that you need to express your anger, it is helpful to strive toward expressing positive feelings as well. Doing so helps to balance the possible negative effects that your anger might have on the problem-solving process. Expressing any genuine feelings of appreciation, respect, or liking of the other party can be helpful in keeping the lines of communication open when angry feelings become particularly intense. Managing anger effectively also requires the ability to receive and accept anger from the other party (Weeks, 1992). Listening to the other person's expressions of anger without retaliating with your own anger and defensiveness can be extremely difficult to do but is a critical skill when trying to keep constructive negotiations and problem solving on track.

In order to resolve conflicts constructively, you need to avoid some of the most common obstacles to conflict resolution. One common obstacle is trying to rush toward a solution too quickly. When the parties hurry to resolve the problem, they tend to take premature, bottom-line stances or put forth pat proposals for solutions that they then hold onto rigidly. One or all individuals can become fixed on a certain position that they believe "must" happen. Taking this stance before thoroughly exploring and understanding the situation, the interests of the various parties, and the range of options that might be available clearly interferes with optimal conflict resolution. Therefore, as you try to resolve conflicts, try to stay as flexible as possible about specific solutions and encourage the other parties to do likewise.

Similarly, trying to find an easy solution through simply "distributing" benefits or resources can seem to be a straightforward and efficient method of resolving conflict. Such approaches, however, tend to block all parties from exploring how the resources or benefits might be expanded. In other words, if the parties involved simply debate about "how to slice the pie," they may overlook possible strategies that might "make the pie larger."

Finally, trying to apply power to solve the problem is another common, unproductive strategy in conflict resolution. The parties involved may try to settle the dispute by focusing on such issues as which party has more support from administrators, which party would win in a court of law, or who has what "rights" in the situation. Though these points hold some validity, such efforts toward solution create animosity and distract the parties from finding reasonable outcomes in which the interests of all parties might be met (Mayer, 1995).

In addition to these general principles, a systematic model for resolving conflicts can be useful in giving direction and structure to this challenging process. Many conflict resolution models have been suggested, with considerable overlap among them. Identifying some common elements in these models, Davidson (2004) suggests a four-step model of conflict resolution consisting of:

1. Developing expectations for win-win solutions. Participants define the issue in terms of their respective underlying concerns, needs, and interests rather than as their "positions." Blaming and criticizing are avoided.

2. Brainstorming creative options. Participants brainstorm alternatives that take into consideration the perspectives of both parties. Ideas are not critiqued or evaluated but emphasis is placed on the quantity and variety of solutions generated.

3. Combining options into win-win solutions. Participants combine ideas from the brainstormed list to create win-win solutions. Win-win solutions are those that meet the needs, interests, and concerns of both parties to the extent possible. Participants can also consider options such as increasing resources, identifying methods of balancing the "costs" of alternatives between the participants, and conceding on matters of less value to the participants. If a solution cannot be found, the problem in doing so is identified and the participants repeat the process.

4. Developing a best alternative into a negotiated agreement. Participants commit to an alternative that is mutually satisfactory and negotiate a clear agreement about how to proceed.

One of the benefits of Davidson's approach is that it has been empirically tested. Participants who had even short periods of conflict resolution training experienced better outcomes in their conflict resolution efforts. Additionally, benefits were found even when only one of the two participants had some training in conflict resolution, although outcomes were better when both participants received training. As you encounter conflicts in the course of your internship, bear in mind that applying basic conflict resolutions strategies can be effective in reducing conflict and stress.

◆ EXERCISE 10.4 SYNTHESIS: LINKING KNOWLEDGE AND EXPERIENCE

Briefly describe a specific conflict that you have dealt with in your internship. Think about the methods of dealing with conflict that you used. (If you have not been involved in a conflict in your internship, you may draw upon an example from your personal life for the purposes of this exercise.)

Now that you have learned about conflict resolution strategies, evaluate the effectiveness of the conflict resolution methods that you used in this situation. Are there specific methods you might have employed to handle this conflict more effectively? In what specific ways might you seek to improve your conflict resolution skills in the future?

Should you experience particularly challenging conflicts during your fieldwork, your field supervisor and/or faculty supervisor should be notified and possibly included in the conflict resolution efforts. Having a third party available as a consultant or mediator in the conflict resolution process can be invaluable.

Developing Positive Self-Talk

It is difficult being a novice, particularly in a setting in which it might seem that all others are uniformly competent and highly experienced while you are the only "new kid on the block." Sometimes you might feel that no one is ever there to witness your shining moments of success but that your deficiencies are glaringly apparent to all. Comparing yourself to established professionals day after day and repeatedly confronting situations that are new, confusing, and even forbidding at times can eventually take its toll on your self-esteem.

A definition of self-esteem by Barker, however, suggests that you can assume direct responsibility for shaping your self-esteem because "an individual's sense of personal worth . . . is derived more from inner thoughts and values than from praise and recognition from others" (1995, p. 340). Another definition, by Long, reflects a similar idea because it suggests that self-esteem is based on "*perceived* [emphasis mine] strengths, attributes, and actions" (1996, p. 178). These statements are consistent with cognitive-behavioral theories, which suggest that our emotions and behaviors are shaped largely by our thoughts rather than by the situations and events that we experience in and of themselves. In other words, it is not necessarily the events in our lives that upset us emotionally, but what and how we think about those events.

According to these theories, one of the most effective ways to change emotions or behaviors is to change the underlying thought process (Beck, 1976; Ellis, 1962). According to the cognitive-behaviorists, consistently thinking thoughts in which you give yourself realistic, accepting, supportive messages is likely to help you feel more confident and relaxed and behave more calmly and appropriately. Realistic, accepting, supportive messages are sometimes referred to as "positive self-talk" (Ellis, 1988). Examples of positive self-talk might include such statements as:

- This is challenging but I can handle it.
- My clients are responsible for making changes in their own lives. I cannot do this for them.
- I will do many things right, but at times I will make mistakes. This is normal and acceptable.
- I do not have to be perfect to do a good job.
- I have learned quite a bit about my profession, but I cannot expect to know everything.
- It is OK to say, "I don't know."

Note that positive self-talk does not consist of unrealistic, overblown statements about your abilities or your performance. Likewise, positive self-talk is not simply "positive thinking" in which you insist in your own mind that all will go well (Meier, 1989). These types of statements tend not to be effective because they are unrealistic—that is, not based on reality. Unrealistic messages are neither believable nor reassuring over the long run.

"Negative self-talk" occurs when people give themselves distorted, overly critical, and unrealistic messages about themselves or their situation. Such messages are thought to compound stress reactions and, in some cases, to cause depression and other forms of more serious emotional distress (Ellis, 1988). Common examples of negative self-talk are:

- I must please everyone all of the time.
- If I make a mistake, I am a failure.
- People must give me what I need at all times.
- I should always be available to the people who need me and know what to do for them.

- This is the worst situation imaginable.
- I'll never get it right.
- That client I saw never came back, and it's all my fault.

As these examples illustrate, negative self-talk tends to take some common forms such as "catastrophizing," jumping to illogical conclusions, overgeneralizing, magnifying or minimizing certain circumstances, personalizing, and all-or-nothing thinking (Beck, Rush, Shaw, & Emery, 1979; Freeman, 1990).

Cognitive-behavioral methods have been found to be effective in helping people manage their emotions more effectively (Granvold, 1995). Your work as a human service professional will no doubt present you with many emotionally challenging situations and events. During your internship and throughout your career, you are advised to work on developing positive self-talk and minimizing negative self-talk as an important and effective tool in managing your emotions and your stress in these situations. Your internship gives you an opportunity to observe your mental habits as you deal with various workplace situations and to work on reprogramming your self-talk as needed.

◆ EXERCISE 10.5 SYNTHESIS: LINKING KNOWLEDGE AND EXPERIENCE

Observe your self-talk over the course of a few days in your internship. What messages do you routinely send yourself about yourself, others, and your situation?

As you review these messages, what examples of positive self-talk do you observe? What examples of negative self-talk?

For each negative self-talk statement, write a more realistic, supportive statement that you can use to replace the negative self-talk.

Changing your self-talk habits will take daily practice and effort. Keeping a list of the positive self-talk statements that you have written in this exercise on your desk or in your journal will serve as a good reminder to practice using them each day.

A Student's Reflections on Self-Talk

If I could start my internship over I would talk more to the other staff members. I eventually did this, but by the time I got warmed up to them, there wasn't much time left. As I look back at my first weeks there, I realize I built up a lot of barriers between myself and them in my own mind. I would often think things like, "They don't like me" or "They don't want me around" or "I'm intruding on her turf." Now that I've gotten to know everyone, I see how off target all of this was. I caused myself a lot of stress and missed some opportunities because of my own wrong-headed beliefs about the people I worked with.

Developing Self-Control

One of the most challenging aspects of managing your emotions may be learning self-control. Although self-control is not directly discussed a great deal in human service literature, Cournoyer emphasizes its importance, saying, "As you grow in self-understanding, you will inevitably recognize a parallel need for self-control. . . . [Y]ou must maturely choose your words and actions in accord with professional purpose, knowledge, values, ethics, and agreed-upon goals for service" (2005, p. 38). All of this is done in service to the principle that it is the client's thoughts, feelings, needs, values, and attitudes, and not our own, that must be primary in the helping relationship.

In order to appreciate fully the value of self-control in managing your feelings effectively, it may be useful to think of it in contrast to its direct opposite, impulsivity. Impulsivity may be defined as "the inclination to act suddenly, in response to inner urges, without thought and with little regard to the consequences of the action" (Barker, 1995, p. 181). The human service intern or professional who speaks or acts impulsively could potentially do a great deal of damage to others as well as to him or herself.

Maintaining self-control and avoiding impulsivity require a great deal of self-monitoring and hard work. A supervisor early in my career shared an observation with me regarding this that I have often found to be true. After an encounter with a particularly frustrating client, he said that in human services work you are often tired at the end of the day, not from what you have done or said, but from what you've worked so hard *not* to do or say. Because such self-restraint is often required, it is essential for workers to find acceptable ways to express these unexpressed feelings and frustrations. Talking with your supervisor or other co-workers about your feelings in such situations can provide a much-needed release of frustration and stress.

A Student's Reflections on Self-Control

There is one particular resident at the group home who is always putting his arms around any woman that comes along, trying to hug and kiss them. The staff has been working with him really hard to change this behavior, but anytime a new woman comes into the picture, it seems that he has to learn it all over again. We are all being very consistent with him, telling him to "shake hands to say hello and use words to tell people you like them." I've gotten pretty good at anticipating his moves and avoiding contact, but at times he slips up on me. At these times I feel extremely angry and my first impulse is to yell at him. I keep telling myself that this young man has serious learning problems and that I must be patient with his slow learning process. "Firmness, consistency, kindness, and patience"—I chant this in my head almost like a mantra at times. He will eventually learn, but my patience is wearing thin!

◆ **EXERCISE 10.6 PERSONAL REFLECTION: OBSERVATION OF SELF AND OTHERS**

Identify at least one instance in which you needed to exercise self-control in handling your emotions during your internship. Describe the situation and how you handled it. How effective were you in maintaining self-control? What did you learn from this experience that can help you in handling similar situations in the future?

Learning to Manage Your Stress

Stress can be experienced as an emotional response and/or a physiological response to a given situation (Robbins, Powers, & Burgess, 1999; Zastrow & Kirst-Ashman, 2004). As an emotional reaction, stress might be experienced mildly as a sense of uneasiness or tension, or more severely as a feeling of fear, dread, or anxiety. As a physiological reaction, stress might be experienced through such symptoms as increased heart rate, increased blood pressure, increased perspiration, and/or increased breath rate.

The situations that trigger these reactions are referred to as stressors. Stressors can be external conditions in the environment, such as a conflictual relationship or a pressing deadline for a project. They can also be internal conditions within yourself, such as your attitudes, beliefs, or expectations (Corey & Corey, 2003). As you think about stress in your own life, keep in mind that a certain amount of stress is thought to be healthy and desirable in that it motivates and stimulates a person to act and achieve. A certain amount of stress is an inevitable part of life. Although some stressors might be reduced or eliminated, it is not realistic to expect that stress will be eliminated entirely. Therefore, although it is always useful to examine how certain stressors might be reduced or eliminated, it is equally important to become skillful in managing stress effectively.

Any aspect of life, even the most pleasant, can cause stress from time to time. Consider, for example, that even your most supportive and caring relationships, such as those with family members and friends, probably create some stress for you sometimes. Certainly some work stress is to be expected as well, especially in a new and demanding situation such as your internship. Although you may be encountering stress in other areas of your life right now, this chapter will focus primarily on understanding and managing work-related stress. The methods of stress management discussed in this chapter, however, can be applied effectively to stress in other areas of your life as well.

Work-related stress is often experienced as an inability to meet the demands of the work environment. You might feel stressed at work from situations such as having too much to do within a limited amount of time, performing a task without sufficient knowledge or preparation, or feeling responsible for an event or situation that seems largely outside your control. Interpersonal relationships at work can also produce stress. Conflict with a co-worker or even constructive criticism from your supervisor might create some stress for you. All of these conditions are considered external stressors, conditions in the environment that trigger a stress response.

While each example noted above involves an environmental stressor, each of these examples might be influenced by an internal stressor as well. You might feel stressed because you expect yourself to get an unrealistic amount of work done within a given time frame or because you expect to know everything about a task before you launch into doing it. These expectations might come from within yourself as much or more than they come from your environment. Similarly, your own beliefs and expectations about relationships might influence your stress level when you are engaged in an interpersonal conflict. For example, you might believe that criticism and interpersonal conflicts are to be avoided and that such events signal some serious shortcoming within yourself. These examples illustrate how stress reactions often involve an intermingling of environmental and attitudinal factors.

Many self-assessment instruments have been developed as self-help tools to assist people in identifying their stress levels and their sources of stress. A particularly useful instrument for your purposes may be the Inventory of College Students' Recent Life Experiences (ICSRLE) (Kohn, Lafreniere, & Gurevich, 1990). The ICSRLE was designed to measure the extent to which an individual has experienced specific stresses, or as the developers refer to them, "hassles," over the past month. This instrument is potentially helpful in identifying specific sources of stress in your life as well as your general level of stress. As you can see, many stressors measured by this instrument are unique to college life.

Assess your own stressors or hassles by completing the Inventory of College Students' Recent Life Experiences (ICSRLE) below. Indicate how much each experience listed has been a part of your life *over the past month*. Mark your answers according to the following guide:

INTENSITY OF EXPERIENCE OVER THE PAST MONTH

0 = not at all part of my life
1 = only slightly part of my life
2 = distinctly part of my life
3 = very much part of my life

_____ 1. Conflicts with boyfriend's/girlfriend's/spouse's family
_____ 2. Being let down or disappointed by friends
_____ 3. Conflict with professor(s)
_____ 4. Social rejection
_____ 5. Too many things to do at once
_____ 6. Being taken for granted
_____ 7. Financial conflicts with family members
_____ 8. Having your trust betrayed by a friend
_____ 9. Separation from people you care about
_____10. Having your contributions overlooked
_____11. Struggling to meet your own academic standards
_____12. Being taken advantage of
_____13. Not enough leisure time
_____14. Struggling to meet the academic standards of others

_____15. A lot of responsibilities
_____16. Dissatisfaction with school
_____17. Decisions about intimate relationship(s)
_____18. Not enough time to meet your obligations
_____19. Dissatisfaction with your mathematical ability
_____20. Important decisions about your future career
_____21. Financial burdens
_____22. Dissatisfaction with your reading ability
_____23. Important decisions about your education
_____24. Loneliness
_____25. Lower grades than you hoped for
_____26. Conflict with teaching assistant(s)
_____27. Not enough time for sleep
_____28. Conflicts with your family
_____29. Heavy demands from extracurricular activities
_____30. Finding courses too demanding
_____31. Conflicts with friends
_____32. Hard effort to get ahead
_____33. Poor health of a friend
_____34. Disliking your studies
_____35. Getting "ripped off" or cheated in the purchase of services
_____36. Social conflicts over smoking
_____37. Difficulties with transportation
_____38. Disliking fellow student(s)
_____39. Conflicts with boyfriend/girlfriend/spouse
_____40. Dissatisfaction with your ability at written expression
_____41. Interruptions of your school work
_____42. Social isolation
_____43. Long waits to get service (e.g., at banks, stores, etc.)
_____44. Being ignored
_____45. Dissatisfaction with your physical appearance
_____46. Finding course(s) uninteresting
_____47. Gossip concerning someone you care about
_____48. Failing to get expected job
_____49. Dissatisfaction with your athletic skills

Scoring the ICSRLE

Add your total points: _____

 ◆ **EXERCISE 10.7 PERSONAL REFLECTION: OBSERVATION OF SELF AND OTHERS**

Focus on two key outcomes from your ICSRLE results: (1) your overall stress level and (2) your particular stressors. Total your responses to each item to find your score on the survey. Scores on the ICSRLE can range from 0 to 147. Higher scores indicate higher levels of exposure to stress.

Now review the items that you rated with either 2 or 3. Higher scored items indicate those stressors are more of an issue for you.

As you review your responses to the survey think about the following: What have been your major stressors over the past month? Which environmental stressors have you experienced? What internal stressors have you experienced? What particular insights do you have about your level of stress as well as your particular stressors?

Once you have identified your stressors, the next question is how to lower your stress or cope with it more effectively. Because stress often involves a combination of both external and internal factors, managing stress often calls for a two-pronged approach—dealing with the environment and dealing with your attitudes. A third response is often helpful as well—simply getting your mind off of the stressful situation and engaging in relaxation-inducing activities (Zastrow & Kirst-Ashman, 2004).

In dealing with real stressors in the environment, skills discussed earlier in this chapter can be very useful. Assertive communication and conflict resolution skills can be applied to many situations in such a way that stress can be reduced or even eliminated. For example, communicating directly with your supervisor if you feel that your workload is too heavy or that you are not sufficiently prepared to carry out a particular task opens the door for necessary adjustments to be made. In this case, the stress response has served a useful function in that it has alerted you to a problem that could and should be solved. Therefore, in dealing with your stress, it is always useful to examine whether the stressful situation can be changed or improved. If so, confronting the stressful situation directly is an excellent coping strategy. A part of stress management, however, involves the recognition that all situations in our environments are not amenable to our influence or control. In such situations, it is wise to recognize these limitations and move on to other coping strategies.

Whether or not the situation is changeable, your attitudes often are. When you are feeling stressed, it is always worthwhile to examine whether your own thoughts, beliefs, or expectations might be creating or amplifying a stress response. An excellent strategy for managing such internal stressors is developing positive and realistic self-talk, discussed earlier in this chapter. For example, you might work on becoming more realistic in your expectations about how much work you can complete within a given unit of time. You might acknowledge to yourself that if you wait until you know everything about a given task before you begin it, you might never do anything in your internship. You might work on reminding yourself that you are in the internship to learn and to be taught, not to prove to others that you already know everything. What all of these examples have in common is that they do not change the circumstances in which you find yourself but they do change the manner in which you perceive, interpret, and think about these circumstances. Stress, in such cases, might have proved helpful to you in the long run by alerting you to your own unrealistic attitudes that were in need of adjustment.

A final but important strategy for dealing with stress involves directing your attention to other matters. Even when you have made your best efforts at reducing stressors in your environment and forming healthy attitudes, you might find yourself continuing to be preoccupied with and troubled by stress. When you know that you have done all that you can to ameliorate the situation and to modify your own attitudes using the methods described earlier, the time has come to work on changing the focus of your attention.

Sometimes a necessary step in changing your focus is venting your concerns to a friend. All of us have had experiences in which we just needed to talk about a situation and our feelings about it in order to get it off our minds. Talking about your experiences and your feelings with a trusted friend can go a long way toward helping you to move on to other matters. In seminars with fieldwork students, I often examine with them their stress-management strategies. I have found that talking with a friend or co-workers is among the most common stress relievers used by students.

A few cautions may be in order, however, in using this strategy. First, although many students use this strategy, they often express concern about and discomfort with the possible confidentiality issues involved. This concern is legitimate and must, of course, be dealt with responsibly in such conversations. Any situations that are in the least questionable in terms of confidentiality should be discussed with your supervisor, faculty liaison, or a co-worker in the agency, rather than with a friend. Many interns also report that they feel most comfortable discussing their work-related stresses with other human service interns who are likely to understand and relate personally to the context of various events. To some extent, field seminars might be used for this purpose.

A second caution about talking out your stress is that talking about the events of the day can simply prolong your attention to these events in a manner that is counterproductive. Although it might be important for you to talk about your experiences and feelings, it is equally important to recognize when to stop. Remember that the point in talking about your concern is ultimately to help you shift your focus away from the stressful events. Continued preoccupation with the stressful events through lengthy discussions is likely only to compound your stress response.

You might think of changing your focus away from the stressful events as a method of maintaining a balance between your personal life and your work life. When focusing your attention away from the stressors, the question becomes, "What shall I focus on?" Each person's answer to this question is highly individualized, drawing upon the individual's supports, habits, hobbies, and interests. Exercise, for example, has been found to be a particularly helpful and constructive method of stress reduction. One person, however, might find more relaxation in taking a walk, whereas another might find playing a vigorous game of tennis more relaxing. Experimenting to find the methods that are most effective for you can be a very pleasurable experience in itself. The following is a list of stress-relieving activities that my students have reported to be helpful to them:

- walking
- bike riding
- swimming
- playing sports, such as tennis, racquetball, golf, basketball, and so on
- running
- working out
- reading (not a textbook!)
- spending time with friends or family
- eating dinner out
- praying
- worshipping or attending religious services
- reading scripture
- meditating
- taking a nap
- sitting outdoors
- going to a movie
- lying in the sun
- working puzzles
- listening to music
- playing an instrument
- gardening
- doing yoga
- taking a long bath
- sky diving (!)

This list is shared not only to give you some possible ideas for your own activities but also to illustrate the breadth and variety of such stress-relieving possibilities. At times, I have also asked students to identify some less-productive methods of relieving stress that they themselves might have used or observed others use. Watching TV for hours, eating junk food, smoking cigarettes, drinking alcohol, and using drugs are commonly cited poor choices for healthy relaxation. This short list clearly illustrates that not all "relaxing" activities are equally healthy, desirable, or productive in the long run. Obviously, making choices that will lead to long-term health and satisfaction rather than seeking the quick fix are advisable.

◆ EXERCISE 10.8 PERSONAL REFLECTION: OBSERVATION OF SELF AND OTHERS

What stress-relieving activities are you engaging in during the course of your internship? Which of these activities, if any, might prove to be particularly unhealthy over the long run? Which ones do you see as particularly healthy and productive? What further ideas do you have about methods you might use to manage your stress more productively?

A Student's Reflections on Stress Management

Gradually I have developed a new evening routine since I have been in my internship. It didn't all happen at once but kind of came together a piece at a time. I find that I really need some time to myself in the evenings, whereas evenings used to be primarily a social time for me. I may go out with my friends at some point in the evening, but I need some wind-down time first. My wind-down time is usually just listening to music in my room. Even just 15 minutes of lying on my bed listening to music with my eyes closed helps me to relax and let go of the day. After that I like to take a walk or do something a little bit physical. More than ever before, I am able to enjoy just being alone. I am a very social person, so this is kind of strange for me. I think I get "peopled out" at work and need a little break before I can be social again.

Conclusion

Learning to manage your feelings and handle stress in ways that are healthy and productive will enhance not only your internship performance but also your performance in your career and your general level of satisfaction in life. Your internship gives you an opportunity to develop more effective skills in these areas. By working on greater self-awareness and self-understanding, you can develop more insight into your emotions and behaviors. Assertive communication and conflict resolution skills are helpful in addressing the various situations that are likely to trigger your more difficult and persistent emotional reactions. Developing positive self-talk and self-control helps you to take responsibility for your own emotional well-being and your own behavior, which can be especially challenging during emotionally intense times. Engaging in relaxation-inducing activities helps you to manage your stress and adds variety and interest to your life. As you learn to manage your day-to-day work stress more effectively, you will be able to enjoy a more balanced life and maintain your energy and enthusiasm for your work.

To learn more about effective stress management, there are many self-help books available as well as excellent websites on the topic. The Mind Tools website is a comprehensive resource offering information about stress management, problem solving, decision making, time management, and many other areas related to professional performance. Also, the International Stress Management Association maintains an excellent website that includes articles about stress, advice about handling stress, lists of useful books and publications, and links to numerous related resources. Finally, the website for Optimal Health Concepts offers a comprehensive list of links related to stress management, including more traditional interventions (such as cognitive-behavioral and relaxation strategies) as well as alternative interventions (such as humor, yoga, and massage). This site also includes a section specifically pertaining to college students' stress.

FOR YOUR E-PORTFOLIO

Describe what this internship has taught you about yourself, your sources of stress, and your ways of handling stress. You might consider the following questions as you develop your thoughts: What have you learned about yourself in this internship? What have you observed about the types of situations that cause stress for you (i.e., external stressors)? What have you learned about ways that you impose stress upon yourself (i.e., internal stressors)? What have you learned about how to reduce and manage these stressors effectively? What particular pitfalls are you learning to avoid as you deal with stress in your life?

References

Alberti, R., & Emmons, M. (1995). *Your perfect right: A guide to assertive living* (7th ed.). San Luis Obispo, CA: Impact.

Barker, R. (1995). *The social work dictionary* (3rd ed.). Silver Spring, MD: National Association of Social Workers.

Beck, A. (1976). *Cognitive therapy and emotional disorders.* New York: International Universities Press.

Beck, A., Rush, A., Shaw, B., & Emery, G. (1979). *Cognitive therapy of depression.* New York: Guilford.

Corey, M., & Corey, G. (2003). *Becoming a helper* (4th ed.). Belmont, CA: Brooks/Cole.

Cournoyer, B. (2005). *The social work skills workbook* (4th ed.). Pacific Grove, CA: Thomson Brooks/Cole.

Davidson, J. (2004). A conflict resolution model. *Theory into Practice, 43*(1), 6–13.

Ellis, A. (1962). *Reason and emotion in psychotherapy.* New York: Lyle Stuart.

Ellis, A. (1988). *How to stubbornly refuse to make yourself miserable about anything—yes, anything!* Secaucus, NJ: Lyle Stuart.

Freeman, A. (1990). Cognitive therapy. In A. Bellack & M. Hersen (Eds.), *Handbook of comparative treatments for adult disorders* (pp. 64–87). New York: Wiley.

Granvold, D. (1995). Cognitive treatment. In R. Edwards & J. Hopps (Eds.), *Encyclopedia of social work* (pp. 525–538). Washington, DC: NASW Press.

Hagen, J. (1999). Conflict resolution: An example of using skills in working with groups. In H. Harris & D. Maloney (Eds.), *Human services: Contemporary issues and trends* (2nd ed., pp. 173–179). Boston: Allyn & Bacon.

Halley, A., Kopp, J., & Austin, M. (1998). *Delivering human services: A learning approach.* New York: Longman.

Johnson, D. (2006). *Reaching out: Interpersonal effectiveness and self-actualization* (9th ed.). Boston: Allyn & Bacon.

Kohn, P., Lafreniere, K., & Gurevich, M. (1990). The inventory of college students' recent life experiences: A decontaminated hassles scale for a special population. *Journal of Behavioral Medicine, 13*(6), 619–630.

Lange, A., & Jakubowski, P. (1976). *Responsible assertive behavior: Cognitive/behavioral procedures for trainers.* Champaign, IL: Research Press.

Long, V. (1996). *Facilitating personal growth in self and others.* Pacific Grove, CA: Brooks/Cole.

Mayer, B. (1995). Conflict resolution. In R. Edwards & J. Hopps (Eds.), *Encyclopedia of social work* (pp. 613–622). Washington, DC: NASW Press.

Meier, S. (1989). *The elements of counseling.* Pacific Grove, CA: Brooks/Cole.

Robbins, G., Powers, D., & Burgess, S. (1999). *A wellness way of life.* New York: McGraw-Hill.

Weeks, D. (1992). *The eight essential steps to conflict resolution.* New York: Putnam.

Zastrow, C., & Kirst-Ashman, K. (2004). *Understanding human behavior and the social environment* (6th ed.). Belmont, CA: Brooks/Cole.

Chapter 11

Trouble-Shooting

All students want to be successful in their fieldwork, and by far most are. Nevertheless, some students do experience particularly challenging situations along the way, which is not surprising given the high expectations that an internship places on students. Consider that as an intern, you must make sound practice decisions, meaningfully apply classroom learning, demonstrate excellent work habits, produce professional quality writing, relate professionally with a variety of staff members and clients, observe and follow professional ethics, and meet the academic expectations of your human service program. Even the most well-intentioned, highly motivated, and best prepared student can hit stumbling blocks on this challenging path.

This chapter is devoted to considering some of the potential stumbling blocks in human services internships and thinking about how to prevent them or how to deal with them should they occur. A risk in discussing such problems is that of raising undue anxiety. To avoid this, there are a few important points that you should keep in mind. Some problems and mistakes are normal parts of life and certainly of student internships. If you expect your performance to be perfect, then you are expecting far too much of yourself. Most internships proceed quite smoothly, and the problems that students encounter are generally minor. Still, as you read this chapter, keep in mind the adage, "Forewarned is forearmed." In keeping with this thought, the discussion here should help you to feel more confident in your ability to avoid any serious problems and to resolve any that might occur.

Making Mistakes and Learning from Them

It is not uncommon for students to worry about making mistakes during their internship. Let's take some of the pressure off by acknowledging that almost anyone starting a new job or field placement is going to make a few mistakes. Most students (and professionals) learn quickly from their mistakes and do not repeat them, and most mistakes are generally not serious and easily reparable.

The first step in learning from a mistake is admitting that you have made one. Sometimes it might not be clear to you whether a particular action was a mistake or not. If you have a gnawing, uncomfortable feeling about something you have done or not done, the best policy is to discuss the issue with your supervisor as soon as possible. In this conversation you can get some direct feedback about the matter and consider whether some course of corrective action is necessary.

At times, however, others may perceive that you have made a mistake when you had no idea that there was a problem. In this case, your supervisor or another worker will point the mistake out to you. In such a situation, you might be caught off guard and feel defensive. Despite this natural response, try to hear the feedback that is being offered and learn from it. Often, all that is needed to settle the waters is a simple and sincere statement such as, "I'm sorry about the mistake. I would like to correct it if I can." Of course, the statement needs to be followed up quickly with whatever corrective action is appropriate.

CASE EXAMPLE 1

Miguel was assigned to a high school guidance office for his internship. Within the first few weeks of his placement, a 14-year-old male student, Antonio, came into the guidance office several times to speak with someone about a conflict with another student. Miguel's supervisor assigned him to talk with Antonio, and the two of them formed a close relationship. In the fourth week of the internship, Antonio walked into the guidance office with Armando, the student he had been complaining about in previous weeks. The two were in a heated conflict. Miguel met with the two together and tried to mediate their conflict.

When the two boys left the office, nothing had been accomplished and Miguel felt angry and frustrated with Armando. As he thought about the conversation, he began to realize that he had spent most of the time listening to Antonio's point of view and at times pressed Armando to change his behavior. He had spent little time hearing Armando's point of view or examining how Antonio's behavior might have been playing into the problem. Miguel realized he had been ineffective in his work with the two boys and was concerned that he might have alienated Armando from ever entering the counseling office again. He thought he should do something about the situation, but what?

◆ EXERCISE 11.1 ANALYSIS

Do you see Miguel's mistake as a reparable mistake? If you were in Miguel's position, what corrective action might you take at this point? What lessons do you think he can learn from this mistake?

Making More Serious Mistakes

Although most mistakes are reparable and educational, occasionally there are mistakes of such seriousness that the student may not be given a second chance. In fact, in some situations, students are terminated from their internship for making a very serious mistake or for an ongoing pattern of unprofessional behavior. Being an intern is similar to being an employee in the agency. Therefore, any behavior that is serious enough to cause an employee to be terminated can result in termination of an intern as well.

In addition to the agency's policies regarding termination, academic human service departments that sponsor internships also have policies about intern behaviors that warrant termination from the internship. These policies are generally published, and students are well informed about them prior to beginning the internship.

According to Wilson, behaviors that are commonly cited as resulting in termination of an intern include physically harming a client, inappropriate behavior that is potentially damaging to "the reputation or functioning of the agency and/or its clients," behavior which "constitutes a danger to the student or others," and "illegal or immoral behavior" (1981, p. 198). Agency policies regarding termination of employees commonly cite such behaviors as unexcused absences and/or tardiness, breaking the law, possessing or using drugs on the job, leaving work without the supervisor's permission, violating safety regulations or policies, violating dress code policies, and insubordination (Truitt, 1991). In some situations, the agency and field supervisor may be somewhat more patient with an intern than they would be with an employee because they realize from the outset that the intern is not highly experienced. Nevertheless, more serious violations can result in immediate termination. Less serious violations are usually dealt with through issuing a warning and allowing an opportunity for performance improvement. Interns should be aware, however, that there are behaviors that can seriously jeopardize their internship.

Although discussion of intern termination may raise anxiety for you as a student, it is helpful to discuss this topic to ensure that you understand the standards and expectations surrounding this issue. Remember that by far most interns excel in their work and are highly valued by their supervisors and other staff members in the agency. Unfortunately, however, some students can at times be naive about the expectations of the internship and not understand how serious the consequences of certain behaviors can be. When serious problems do occur and a student is terminated, it is an extremely difficult and painful situation for all concerned—the student, the supervisor, and the faculty liaison. Prevention of such a scenario is strongly preferred by all parties.

CASE EXAMPLE 2

Bradley was delighted to be placed in the Student Services Office at a local community college for his internship. His field supervisor was the director of student activities. Bradley was responsible for planning and organizing special events. On one particular occasion, Bradley, under the direction of his supervisor, planned a trip to a nearby town for a concert. Students responded well to the idea, and 25 students signed up for the event within just a few days. On the day of the concert, Bradley, a student worker, and the 25 students loaded into vans and attended the concert. As the group drove back, they stopped at a restaurant near the campus for a late dinner. Bradley was of legal drinking age, and he decided to have a beer with his meal.

The policy of the Student Services Office and of the college was that no alcohol was allowed at functions sponsored by Student Services. Also Student Services staff members were expected to respect and enforce the state laws regarding the legal drinking age and to encourage students to respect and obey these laws as well. The philosophy of the Student Services Office also strongly emphasized a clear commitment to providing forms of entertainment and recreation that did not involve alcohol. In addition, Bradley and all of the other human service interns had signed a contract with the academic department that confirmed an understanding and agreement that neither alcohol nor drugs were to be consumed during the internship.

As Bradley and the group enjoyed their meal, a Student Services staff member happened to walk into the restaurant, and she saw Bradley drinking. The next morning she informed his field supervisor of what she had observed. After discussing the situation with Bradley and establishing the facts as previously described, the field supervisor and faculty liaison terminated Bradley from the internship.

 ◆ **EXERCISE 11.2 ANALYSIS**

Imagine that you are Bradley's supervisor or faculty liaison explaining to him the grounds for his termination. What specific grounds for termination would you cite? What risks, if any, might the Student Services Office and the Human Services Department have incurred if they had allowed Bradley to continue in his internship following the incident described? What lessons can he learn from this situation?

The difficulties that most students encounter in the internship are generally much less serious than Bradley's situation. Nevertheless, even far less serious violations of supervisory expectations can harm how a student is viewed by the supervisor and can indicate significant changes that the student needs to make before assuming the full responsibilities of a professional position.

Maintaining Good Work Habits

Ironically, after years of preparation for entering the professional world, the issues that most often compromise a student's performance in the internship are generally related to simple, basic work habits, not the finer points of academic preparation or professionalism. Although a student might have excellent helping skills and impressive theoretical knowledge, these will likely go unnoticed and unappreciated if the intern is not considered a good worker. In the working world, there are some generally recognized commonsense traits of a good worker in virtually any environment. These traits include

- arriving at work on time;
- staying at work for the full day;
- coming to work early and/or staying late occasionally, if this is needed;
- attending work reliably;
- using good grammar;
- dressing appropriately for the position;
- using time productively;
- being pleasant, energetic, and alert;
- relating cordially to co-workers;
- writing well;
- being flexible; and
- carrying out responsibilities reliably, promptly, and efficiently.

Supervisors value these behaviors in their employees in any type of organization, and human service organizations are no exception. Although you are probably not being paid for your internship, you should treat it like a very desirable job. If you are about to complete your academic program, your supervisor will be evaluating your performance in relation to your readiness to handle an entry-level position in the organization. It should be your goal to demonstrate that you are fully prepared to assume such a position and handle it well. If you enter your internship consistently demonstrating the behaviors above, you are well on your way to at least a satisfactory performance, possibly even an outstanding one.

CASE EXAMPLE 3

Tamara, a senior human service major, performed her internship in a vocational rehabilitation program. Her field supervisor noticed that Tamara arrived 15–20 minutes late the first three days of her internship. Having allowed for an adjustment period, the supervisor mentioned the tardiness on the third day and explained to Tamara that it was important for her to arrive on time, especially because this was a behavior much emphasized with the clients in the program. Tamara apologized and explained that tardiness had always been a problem for her and that it was in no way a reflection of her interest in or commitment to the internship. There was little to no change, however, in Tamara's arrival time in the days following this conversation.

A week later, as the problem continued, the field supervisor brought up the problem in a meeting that they had with Tamara's faculty liaison from her college. The supervisor explained in more detail the various reasons why promptness was important in the organization. In this same meeting, Tamara's faculty liaison also emphasized that promptness and reliability were important expectations of the department and stated her expectation that Tamara would improve her performance in this area during her internship. Nevertheless, the tardiness continued even after this lengthy discussion of the problem.

Tamara at times did excellent work within the agency. She worked well with clients and completed several administrative projects. She had used her computer skills to create a highly effective brochure for the agency and produced a statistical report that was long overdue within the agency. She was praised for her good work and enjoyed a pleasant relationship with her supervisor. Tamara felt that she was working hard in her placement and that she was doing an excellent job.

In her evaluation conference, her supervisor praised her for the strengths she had demonstrated and the work she had performed for the organization. She evaluated her skills highly in certain areas. Tamara was crushed, however, when her supervisor responded with a "No" to the question, "If a position were available within your agency for which this student qualifies, would you recommend her as a candidate for the job?" Her supervisor explained that due to Tamara's habitual tardiness, she would not hire her and could not recommend her for a job within any agency. Tamara was disappointed in her evaluation and phoned her faculty liaison to complain about her "unfair evaluation." She stated that there were employees who came to work late and that she was being singled out unfairly for this offense.

 ◆ **EXERCISE 11.3 ANALYSIS**

What are your thoughts about Tamara's evaluation and her reaction to it? If you were the faculty liaison responding to Tamara's concern about her evaluation, what might you say?

Trouble-Shooting

Maintaining Appropriate Professional Boundaries

A frequently cited challenge for human service interns is learning to maintain appropriate professional boundaries (Alle-Corliss & Alle-Corliss, 1998; Royse, Dhooper, & Rompf, 1999; Sweitzer & King, 2004; Thomlison, Rogers, Collins, & Grinnell, 1996). Professional boundaries have to do with making sound decisions about what behaviors are and are not prudent within your professional role with clients and colleagues. Boundary issues are involved in questions such as:

- ◆ "How much should I tell clients about my own life?"
- ◆ "Is it alright to socialize with clients?"
- ◆ "How involved should I get in the lives of my clients?"
- ◆ "Is it appropriate to accept a date with an employee in my field agency?"
- ◆ "Is it appropriate to develop a personal friendship with my supervisor?"

There is evidence that suggests that most human service interns need more training and guidance about such professional boundary issues than they generally receive in their educational programs (Slimp & Burian, 1994).

A common boundary issue that human service interns struggle with is the urge to "rescue" clients. Because human service professionals want to help others, it is easy at times to get overinvolved in clients' lives in an effort to rescue them from the distress that they are experiencing (Alle-Corliss & Alle-Corliss, 1999). Efforts to rescue might range from offering false reassurance in an effort to decrease a client's pain at the moment to more extreme urges to take clients home with you or to fight their battles for them in their conflicts with others. Whether mild or extreme, efforts to rescue clients are almost always counterproductive because they rob clients of the opportunity to grow and learn through grappling with their own problems. Clients instead tend either to become dependent upon the worker to handle their problems in the future or to become angry at the worker for not rescuing them successfully. In either case, both the client and the worker suffer in the long run.

In relation to questions about professional boundaries, it is generally best to "err on the side of caution." Some case examples illustrate and clarify the nature of professional boundary issues.

CASE EXAMPLE 4

Larry worked in a group home for teenagers with emotional and behavioral problems. One evening while he was sitting in his apartment, the doorbell rang and he found two of the group home residents standing at his door. One of the boys, Carlos, had asked him earlier in the week where he lived. Although he initially hesitated, Larry had told Carlos the name of his apartment complex. When Carlos had asked which apartment he lived in, Larry was afraid it would seem rude not to tell him, so he had quickly blurted out the apartment number, thinking, "Oh, what can it hurt?" As Larry stood at the door looking at the two youths, he remembered and deeply regretted that conversation, but he let them in.

The boys informed him that they had run away from the group home and did not intend to go back. They begged him not to call the group home staff and related stories about how harshly they were treated there. They stated repeatedly that Larry was "the only one we can trust, the only one we can talk to." Larry spent hours with the boys trying to convince them to go back to the group home voluntarily. As time passed, he grew increasingly uncomfortable with his involvement in the situation. Convinced at last that the boys could not be persuaded to return on their own, Larry finally called the group home to inform them of the boys' whereabouts. Unfortunately, 6 hours had elapsed since the boys had left the group home. Staff were upset and worried, and the police had been notified. Larry faced possible disciplinary action within his field agency, possibly even termination.

 ◆ **EXERCISE 11.4 ANALYSIS**

At what points did Larry have the opportunity to handle this situation more productively?

What professional boundary issues were involved in this situation and how should Larry have handled them?

What are your thoughts about what might have motivated Larry to handle the situation in the manner that he did?

What lessons can Larry learn from this situation?

Boundary issues can also become involved in interns' relationships with agency staff. Interns have at times been asked out by co-workers, asked to buy expensive products that staff members were selling, and even been employed to work in businesses owned by their field supervisor during the time frame of their internship. Although such situations usually do not carry the risk of potential harm to clients, they can become quite troublesome for the student, potentially damaging the quality of the student's educational experience as well as the student's relationships with the supervisor and/or co-workers. Should a staff member approach you with any request that seems questionable, inform him or her that you would like some time to think about it before responding, then discuss the situation with your supervisor and/or faculty liaison.

CASE EXAMPLE 5

In the first week of her internship, Tracy learned that a co-worker in her agency was running for the local school board. The co-worker, Mary, asked Tracy if she would be willing to help with her campaign. Tracy liked Mary and had difficulty saying "No" to others, so she agreed to help. Never having been involved in politics, she rationalized that working on a campaign would be educational for her. Because the campaign work was to be done outside the internship and outside the business hours of the agency, it did not occur to Tracy that it could cause problems in her internship.

After two weeks of helping with the campaign and learning more about Mary's positions on issues, Tracy was not sure that she really felt comfortable supporting her as a candidate. She tried to pull out of her commitment subtly and gracefully by explaining that she needed to spend more time on her school assignments. She noticed a cooling of Mary's attitude toward her but decided that this was to be expected.

Tracy began to notice that Mary, as well as other workers, were not asking for help with their work or assigning her tasks as they did in the first weeks of her internship. She felt mildly uncomfortable about the situation but was not particularly worried or concerned about it until a supervisory conference in which her supervisor addressed the issue. Her supervisor said, "I'm concerned that you might be developing a reputation among the staff for being unreliable. I suggested to Mary that she get your help on a project she is doing, and she said that you couldn't be counted on to follow through. Do you have any idea about where this impression might have come from?"

 ◆ **EXERCISE 11.5 ANALYSIS**

What steps can Tracy take to correct the situation in which she finds herself?

What steps might she have taken earlier to avoid the situation?

What lessons do you think Tracy needs to learn from this situation?

Guarding Against Dual Relationships

The case of Tracy involves a dual relationship with a co-worker. Dual relationships constitute a particular type of boundary issue that is a much-discussed and controversial issue in human services today. Due to the frequency and complexity of this particular problem, it is worthy of special attention here. Dual relationships occur when a helping professional assumes a second role with a client or colleague in addition to the professional role (Barker, 1996). When such relationships occur with clients, they are clearly unethical according to the Ethical Standards of the National Organization for Human Services (see the Appendix). Given the clarity of this point, it would seem that professionals should not be confused as to whether it is appropriate to develop a personal friendship, social relationship, or business association with a client. By the same token, it is considered inappropriate for professionals to provide professional services for friends, family members, and other individuals with whom they have a preexisting relationship. Despite the clarity of these guidelines, the real world can have a way of blurring this boundary. For example, if you are only superficially acquainted with someone, is it acceptable to work with them professionally? If a client brings you a small birthday gift, does this mark the beginning of a social relationship? If you know a close family member of the client personally but not the client himself, does this create a boundary violation? Due to the changing nature of relationships and circumstances over time and gray areas like those in the previous examples, workers must be constantly vigilant about the nature of their roles with the clients they serve.

CASE EXAMPLE 6

Mingyu worked as a gymnastics coach at the local YMCA while attending college. There he worked with Don, a talented 12-year-old who was on the YMCA's gymnastics team. While performing his internship as a Juvenile Court counselor, Mingyu was shocked to see Don in juvenile court due to charges of vandalism and theft. Although the situation made Mingyu somewhat uncomfortable, he did not feel there was any real conflict because Don was assigned to a different court counselor. Within a few days, a number of issues began to complicate the picture. For example, in staffing later that week, information was shared about Don's family that Mingyu felt sure Don's family would not want him to know. Also, certain information was shared about the family that Mingyu knew for a fact was false. He suspected that either Don, his parents, or both had lied to his court counselor. As the staffing came to a close, Mingyu's heart sank further when the decision was made to refer Don to a group that Mingyu and his supervisor were co-leading. Recognizing the many dual role issues involved, Mingyu scheduled a meeting with his supervisor immediately to discuss the situation.

 ◆ **EXERCISE 11.6 ANALYSIS**

What do you see as Mingyu's options in resolving this problem? If you were in his position, how would you resolve the problem?

How should Mingyu handle the conflict between what he knows about Don's family and the information reported by the court counselor?

What role, if any, should he have in any discussions about Don that might occur in the agency?

Fortunately, Mingyu was wise enough to recognize the difficult issues raised by the situation in which he found himself. Talking with his supervisor was exactly the right thing to do. Even his supervisor, however, felt that he needed some guidance in the process of sorting the situation out, prompting a wider discussion of the issue among the staff as a whole. In the long run, Mingyu's raising his concerns brought the entire agency to a higher level of awareness about the potential for dual relationship issues within their work.

Although the ethical problems in dual relationships were first discussed in human service literature in relation to worker–client relationships, the discussion has in recent years expanded to include concerns about other types of dual roles during internships, particularly between interns and supervisors (Bernard & Goodyear, 1992; Keith-Spiegel & Koocher, 1985; Slimp & Burian, 1994). When supervisors and students develop roles and relationships with one another in addition to their supervisory relationship, complications can occur that compromise student learning and in some cases might even result in exploitation of the student.

Some common secondary roles that have been discussed in the human service literature are sexual relationships, social relationships, and business relationships (Slimp & Burian, 1994). Kaiser shares an example of a student whose learning was harmed by a dual social role with her supervisor. Several years after her field experience, the student says,

> [W]e had some difficulties in confronting each other about things. I think maybe we had a secret pact to not to talk about some things. There were times when we did talk about uncomfortable issues. I believe we only went so far, though. I think our social relationship got in the way. (Kaiser, 1997, p. 66)

In this particular case, the student was not seriously harmed by the dual relationship nor was she particularly concerned about it at the time. In retrospect, however, she realized that her friendship with her supervisor had created a less-than-ideal learning environment for her. As Kaiser (1997) points out, a dual relationship with a supervisor is problematic because it creates a competing agenda in the supervisory relationship. Once an additional agenda is introduced, there is the risk that the goals of that agenda become primary, while the goals of supervision and student learning become secondary.

Social relationships between supervisors and students can cause harm, but the problems created by the development of sexual relationships, therapeutic relationships, and business relationships are generally seen as even more potentially damaging. Kitchener (1988) suggests that dual relationships are more problematic when (1) the two relationships have very different expectations, (2) the obligations of the two relationships are very divergent, and/or (3) there is a wide differential in power and prestige between the two individuals. While a social relationship may be somewhat similar to the supervisory relationship, sexual relations, psychotherapy, and business dealings have markedly different expectations and obligations. Therefore, according to Kitchener's framework, one would expect the risks of harm in such dual relationships to be greater. Also, since there is a considerable power differential between students and their supervisors, Kitchener's model would suggest that there is, in all cases of dual relationships between students and supervisors, a significant risk of students being harmed. The following case example illustrates how a dual relationship involving a business deal between an intern and her supervisor harmed the quality of the intern's experience.

CASE EXAMPLE 7

Elena had been searching for a much-needed larger house for months prior to beginning her internship at the local homeless shelter. After her internship began, Elena spoke with co-workers often about various houses she had seen with her realtor. Disappointment and frustration mounted as the weeks rolled by and a satisfactory house was yet to be found.

Meanwhile, Elena's internship was going well. She was learning a tremendous amount from her supervisor, Grace, an older woman with extensive experience whom Elena admired enormously. On two occasions she had visited Grace's home, once for a staff party and once to drop off a report. On both occasions Elena had admired Grace's home and had even commented, "If only we could find a place like this." Therefore, Elena was delighted when Grace announced that she and her husband planned to sell their house and move into a townhouse. Elena immediately and enthusiastically indicated her interest in the house. Grace had mixed feelings about the situation. She knew that there could be trouble in doing business with her student, but it pleased her to think about Elena and her family enjoying the house. Also, she knew what a difficult time Elena had been having in finding a suitable place. Within days, Elena contacted Grace's realtor and viewed the house with her family. Shortly thereafter, she made an offer on the house. The offer was a good one financially, and Grace, with some misgivings, decided to accept it.

Over the next few weeks, Elena and Grace increasingly spent supervision time, as well as any other free moments at work, talking about the house and the details of the sale. Each "supervisory session" ended with the mutual promise that next time they would stay on task, but each week there seemed to be more pressing issues to deal with about the house. Matters grew worse when conflicts arose between them about the conditions of the sale. Elena began to request repairs that Grace felt were unreasonable. Elena felt put out by Grace's asking for an extension on the closing date because construction on the townhouse that she and her husband were building was running behind. Tension and mutual mistrust began to develop, and both Grace and Elena began to avoid each other as much as possible. By the end of the placement, Grace and Elena were spending limited time with one another, and Elena was informally seeking supervision on a "catch-as-catch-can" basis from other staff members.

◆ **EXERCISE 11.7 ANALYSIS**

At what point in the supervisory relationship did a clear dual role develop between Elena and Grace?

What responsibility did each individual bear in ensuring that the integrity of the supervisory relationship was preserved?

How might Grace and Elena have handled the situation differently in order to avoid the dual relationship?

Maintaining a Balance Between Your Work Life and Your Personal Life

The students in Case Examples 4 through 7 had difficulty maintaining appropriate boundaries in their professional relationships. Another type of boundary issue that interns sometimes struggle with is the boundary between their work lives and their personal lives.

In dealing with the demands of the internship, students can find it difficult to maintain the right balance between their work responsibilities and their personal lives. The typical day of the college student usually includes a few classes with some breaks during the day for running errands, doing chores, spending time with friends or family, studying, or just simply resting. The internship, like a job, does not offer this flexibility. Students must figure out how to reorganize their priorities and their time to allow for what is usually at least an 8-hour-a-day job. Because the internship can at times be stressful and demanding, it is also a particularly important time for students to take care of themselves, physically and emotionally. Finding the right balance between the demands of a career and the demands of one's personal life involves a lifelong process of learning and self-discipline for many human service professionals. The internship is an excellent time to begin cultivating good habits in managing these boundaries.

A Student's Reflections on Managing Boundaries Between Work and Personal Life

There are things that I have practically considered it my God-given right to do every day that I have just had to give up. Talking on the phone to friends, watching TV, and even reading the newspaper have become luxuries occasionally indulged in, not everyday occurrences. Although doing this internship has been a real stretch for my schedule, it has forced me to sort out what my priorities are. My evenings I devote pretty much to family until the children are in bed. I do manage to do a little reading and such occasionally while they are still up, especially if they have lots of homework, too. My husband has taken over some of the errands and such that I used to do during the day. Occasionally, I might try to do a couple of things during lunch, but otherwise I keep my day exclusively focused on my internship. My goal after I graduate is to get a full-time job, so I see this as way to start making that transition. But I won't pretend that it's easy!

Although most students strike a healthy balance between their internship and their personal lives, some are swept into a pattern of working far too much. These students might consistently stay at the office late, arrive early, and take work home with them. Even when they are "off," they may still be working mentally, thinking about problems at work or planning the next day in their minds. Because such hard-working, highly dedicated behavior is often praised and rewarded in our society, the student who succumbs to this pattern can easily feel that this is the right thing to do. Through their own work habits, supervisors and co-workers might even covertly and unintentionally communicate that this level of time commitment is expected from others.

Although you might initially feel good about yourself for working so hard, this working style can potentially lead to negative consequences over the long run. Human service professionals must practice good self-care in order to maintain their energy. Personal time is necessary to take care of yourself physically, emotionally, socially, and spiritually. Even so, taking time out for even the most basic essentials of self-care, such as sufficient sleep and nutritious meals, is too often overlooked by busy interns. Such self-care habits are absolutely necessary, however, because without them you will find it difficult to maintain the energy required for a demanding career (Littleton, 1996).

Students who begin their careers working excessively may quickly begin to feel the effects of burnout (Russo, 1993). Burnout occurs among many human services professionals who do not learn to manage the demands of their profession effectively (Cherniss, 1980; Farber, 1983; Freudenberger, 1983). Burnout has been defined as "a state of physical, emotional, and mental exhaustion that results from constant or repeated emotional pressure associated with an intense, long-term involvement with people. It is characterized by feelings of helplessness and hopelessness and by a negative view of self and negative attitudes toward work, life, and other people" (Corey & Corey, 2003, p. 352). The symptoms of burnout among human service professionals include physical and emotional exhaustion, irritability, negativity, decreased productivity, and feelings of detachment and callousness (Sweitzer, 2004). At times, burnout can even lead to more serious symptoms of depression and anxiety and physical symptoms such as headaches and stomach upsets (Schram & Mandell, 2006). Beyond the personal level, agencies' performance and effectiveness can also be adversely affected by burnout within the staff, thus harming the quality of services to clients (Alle-Corliss & Alle-Corliss, 1998).

Obviously, the consequences of burnout are potentially serious. Therefore, you should strive to develop habits during your internship that will help you to avoid burnout. This is important not only to maintain the quality of your life, professionally and personally, but also to assure the quality of services to clients.

CASE EXAMPLE 8

Van was fortunate to work during his internship under the supervision of a professional whom he genuinely admired. Mr. Stevens, Van's supervisor, was the director of a crisis shelter for runaway teens in a major metropolitan area. He had started the shelter himself and had nurtured it from a struggling, poorly funded, poorly staffed program to a stable, professional, and highly respected organization. Moreover, he had stayed directly involved with the young people who were the clients of the shelter, rather than moving into doing purely administrative work as the agency had grown. Mr. Stevens truly cared about the kids in the program and routinely worked 12–16 hour days. He spent hours each day listening to the teens, counseling them, and helping them make important decisions about their lives. In addition, he was responsible for and effectively carried out a myriad of administrative responsibilities.

In the beginning days of Van's internship, he wanted to be at the shelter every minute. He felt that if he left, he was going to miss something. There was so much he had to learn, from Mr. Stevens as well as from the teens, that he wanted to be there and use every moment wisely. By the end of the first week of his internship, Van had worked 60 hours. He worked a similar schedule during the second and third weeks. By the end of the fourth week of the internship, he was beginning to feel pretty tired. He knew that he needed to establish a different schedule for himself but found it a difficult topic to bring up with Mr. Stevens, especially since Mr. Stevens seemed to be immune to fatigue himself.

During the fourth week of the internship, Van's faculty liaison pointed out to him that he had turned in two assignments late and still had two assignments that he had not turned in at all. Van became very angry and defensive. He argued that the human service department expected far too much of its interns and that there was no way he could live up to everyone's expectation of him. He expressed resentment toward having "all these stupid assignments" and added that he had not even had a conversation with his own roommates or been out with his girlfriend in over two weeks. In closing he stated, "I'm beginning to think I'd be better off as a shoe salesman or something. You don't have to feel guilty at 5:00 if all you're leaving behind is a bunch of shoes."

EXERCISE 11.8 ANALYSIS

What incentives are there for Van to change his work habits? What forces are operating to keep him "stuck" in his current behavior patterns? What aspects of Van's situation can you personally identify with? How have you dealt with such boundary issues in your own internship?

In order to avoid overworking and burnout, try to find ways that you can take care of yourself daily. If you are struggling with this issue, review the section "Learning to Manage Your Stress" in Chapter 10. A number of ideas are discussed there that you might find helpful.

A Student's Reflections on Stress and Personal Boundaries

Yesterday I had my first exposure to intake. I helped do intakes all afternoon. It was basically 4 hours of dealing with the most depressing human pain, misery, and poverty that I have ever encountered. It was tough to enjoy my nice comfortable home and cozy gas heat when I got home knowing about all those people who can't pay their gas bills and don't have anything. I thought about it all evening and even dreamed about it when I finally went to sleep. I confessed this rather casually to my supervisor this morning. Upon hearing about my night, she shut her office door, sat me down, put her hands on my shoulders and said firmly but quietly and kindly, "You can't take this home with you. If we all did that, we would go insane and we wouldn't be any help to anybody." I appreciated her way of putting this. To help others I must let go of the misery, including my own guilt, and take care of myself. Looking at it in this way will help me exercise the discipline I need in order to avoid getting too involved in my work.

In contrast to those students who have trouble keeping their internship within its proper boundaries, other students might have the opposite problem. These students have difficulty keeping their personal lives within proper boundaries. This might be the case because they are completing their internship at a particularly difficult time in their personal lives. If so, these students might come to the internship worried about their personal problems. Very practical aspects of some types of problems can also directly interfere with the internship's demands. For example, even ordinary tasks, such as arriving on time, might be a challenge for the student who has unreliable child care or a sick relative to care for. Students in extremely demanding personal situations during their internships may be well advised to seek some counseling during this time. If you should find that your personal situation is too demanding, it may be necessary to withdraw from the internship. In most situations, however, this extreme is not warranted. Throughout a human service professional's career, conflicts between personal issues and career have to be managed. Eventually, all responsible workers must learn to do this effectively. Although it may not be ideal to have to begin learning this difficult lesson during your internship, it is virtually inevitable that you will confront that challenge sooner or later. You might look upon the internship as an opportunity to start this learning.

Many students struggle with keeping their personal lives contained during the internship, not because of any particular personal problems they might be having, but due to limited coping skills in dealing with everyday concerns. These students might benefit from such relatively straightforward measures as improving their organizational skills, learning time-management skills, or developing greater assertiveness in their personal relationships.

CASE EXAMPLE 9

Lea was the kind of person whose friends always came to her for help. In fact, it was this observation about herself that had led her into considering a human services career. At times, she wondered if she attracted "needy people." It seemed there was always someone who needed a ride someplace, a friend who was in a crisis and needed to talk, or a family conflict that she needed to mediate. Although Lea had never been the strongest student in the classroom, she knew that she would excel in her internship. She was a helper by nature and was willing to do anything to help anyone. She was looking forward to her internship because she knew she would succeed.

Two weeks into her internship, after she had fallen asleep in a staff meeting, her supervisor pulled her aside to talk. She pointed out that Lea had come in late on two mornings due to oversleeping, had left early on Friday to take a friend to the airport, and had often seemed to have low energy during the work day. Among her supervisor's other comments, she said, "It's beginning to seem that this internship isn't very important to you."

Lea protested that this was not the case. She explained that her roommate was having some problems with her boyfriend, and that consequently, she and her roommate were up talking until the wee hours of the morning almost every night. She explained that she simply had not been getting enough sleep and that she was indeed very serious about her internship.

Lea's supervisor communicated clearly to her that her performance needed to improve in the areas they had discussed. Lea left the conversation disappointed in herself but also feeling somewhat angry and misunderstood. After all, she was just being a good friend to her roommate and had been about the business of helping someone who needed it. Isn't that what human services is all about? Why, she wondered, couldn't her supervisor just understand that and give her a break?

 ◆ **EXERCISE 11.9 ANALYSIS**

Based on the information in the case example, what understanding does Lea seem to have about what "helping" is?

In order to change her own behavior, Lea might need to rethink her ideas about helping. How might the concept of personal boundaries be useful to her in this regard?

What skills and/or behaviors does she need to apply in her personal relationships in order to establish better boundaries between her professional life and her personal life?

Much of what beginning human service professionals learn in an internship has to do with boundary issues. The internship marks the beginning of what is often a long-term challenge to find the proper balance and boundaries between your personal life and your professional life. Finding this balance is an important element in learning to take care of yourself as a professional. Another element in learning to take care of yourself as a professional involves maintaining physical safety while on the job.

◆ EXERCISE 11.10 PERSONAL REFLECTION: OBSERVATION OF SELF AND OTHERS

To what extent are you maintaining a healthy balance between your internship and your personal life? In what ways, if any, has this been a challenge for you? In what ways do you need to improve your self-care in order to be at your best for your internship? What kinds of interpersonal boundary issues have been most challenging for you during your internship? How have you handled these situations?

Keeping Safe

Unfortunately, acts of violence have increased in society in recent years. Consistent with this disturbing trend, incidents involving physical aggression against human service workers have also increased (Newhill, 1992, 1995; Vallianatos, 2001). Research suggests that certain populations that human service professionals serve tend to be associated with a higher probability of aggressive behavior than are others. One of the best predictors of violent behavior is previous episodes of violent behavior (Star, 1984). Newhill, for example, found that "being young and male, having a history of substance abuse and weapons possession or a criminal record, and having a history of violent behavior [were] associated empirically with a higher probability of being violent in the future" (1995, p. 633). Also, certain symptoms of mental illness, such as paranoid delusions and command hallucinations (i.e., hallucinations in which the individual is ordered to perform certain acts), have a higher association with violent behavior (Newhill, 1992).

Beyond certain client features, workplace conditions can also create increased vulnerability to violence. Budget cuts and decreased staffing in certain environments, for example, have been found to increase the risk of violent incidents (Hiratsuka, 1988; Petrie, Lawson, & Hollender, 1982; Schulz, 1987, 1989). Schulz (1987, 1989) found that violence against workers was common within certain types of settings, such as corrections, health and mental health, and services for people with disabilities.

Such information suggests that you should be vigilant in gathering information from your supervisor about potential risks that might exist in your particular setting as well as information about how to minimize and deal with these risks. Knowing as much as possible about the populations that your agency serves, as well as its security and safety measures, will help you to take whatever preventive measures are appropriate and prudent within your setting. Ask your supervisor about any aggressive incidents that might have occurred in the past. Also inquire about any staff training or operating policies that are in place to protect workers. In some settings in which there is a particularly high risk of clients becoming aggressive or out of control, workers are trained in specialized methods, such as use of physical restraints. Also, some agencies flag certain cases as high risk by placing a color sticker or other marker on the client's record. This serves to alert workers to use special precautions when working with this client (Griffin, 1995).

Although not all agencies provide formal training for workers on dealing with aggressive clients or flag high-risk cases, many do routinely advise workers on certain practices that can minimize their vulnerability. Workers may be advised, for example, to arrange their offices such that they are seated between the client and the door. Often, agencies advise workers to perform certain types of potentially high-risk tasks, such as home visits, in pairs. Some agencies even encourage workers to leave their office doors slightly ajar when meeting with clients in the agency. This enables nearby workers to be at least somewhat aware of activities within the office and makes it easier for the worker who is in the office to summon help or leave the room.

As a student, you should be reassured by the fact that supervisors usually avoid intentionally involving students in high-risk situations. In some settings, however, a certain amount of risk is ever present and unavoidable. Also, the unexpected can occur in any setting. Therefore, it is wise to prepare yourself for the possibility of dealing with an aggressive client, even if such an event seems unlikely.

If, despite everyone's best efforts at prevention, you should find yourself in the presence of a client whose behavior seems threatening, certain behaviors on your part can help to deescalate the situation. Try to stay calm, speaking slowly in a normal conversational tone. Be respectful toward the client and listen empathetically to his or her feelings. Avoid arguing with the client or responding angrily (Royse et al., 1996). When responding to the client verbally, keep in mind that either passivity or coerciveness on your part may result in the client becoming more upset. Assertive communication can be effective in setting appropriate limits and providing a much-needed sense of structure and safety for a client who is feeling out of control. Even so, take care not to come across as controlling or demanding as this will likely intensify the client's anger and anxiety (Murdach, 1993; Weinger, 2001). Weinger (2001) suggests helping the client talk out angry feelings rather than act them out. Try to help the client regain a sense of control and status in the conversation through tactics such as speaking simply, breaking down seemingly overwhelming problems into manageable steps, or redirecting the client away from the anger-provoking topic (Weinger, 2001). If you feel that the situation is imminently dangerous, make an excuse to leave the room. Offer to get a drink for the client or say that you need to get something for yourself (Royse et al., 1996). Once you have left the room, alert other workers immediately so that the client is not left unsupervised (Faiver, Eisengart, & Colonna, 1995).

Nonverbal communication is, of course, at least as important as verbal communication in dealing with aggressive clients. Do not touch the client. Eye contact that conveys your involvement with the client is positive but avoid making eye contact that is so intense that it might be interpreted as aggressive (Weinger, 2001). Give the client more than the usual amount of personal space, if possible. Avoid making sudden movements and explain whatever movements you need to make. Making an explanatory statement, such as, "I'm going to pick up this folder now so I can look at my notes" can help prevent an agitated client from becoming even more alarmed, suspicious, or upset (Shea, 1988). Sit down while speaking with the client and encourage the client to sit down as well. Standing is generally interpreted as a more aggressive physical stance (Royse et al., 1996). You might also consider ways to make it easy for the client to leave in these situations, thus allowing the client the option of "flight" over "fight" (Weinger, 2001).

Finally, use common sense and trust your instincts in such situations. When going into situations that seem risky, have someone come along with you. This is a particularly good idea when making home visits or venturing into high-crime areas. Also, keep in mind that in particularly high-risk situations, law enforcement officers can be asked to accompany workers. Do not hesitate to ask for support if you are feeling uneasy about a situation. Workers who have been involved in violent episodes frequently report after the event that they had a gut-level feeling that the situation could be dangerous. In such situations, even if there has been no real or direct threat, it is important to tell someone and take precautionary measures (Safety First, 1998).

Unfortunately, rather than asking for help in such situations, workers too often proceed alone, feeling too embarrassed to ask for assistance or even to admit that they are fearful (Newhill, 1995). Moreover, even professionals who have been physically harmed by clients are often reluctant to report it due to shame and fears that they might be perceived as unprofessional (Weinger, 2001). As with some other topics that have been discussed in this book, it is better to "err on the side of caution." Asking for support when you need it is not a sign of weakness or failure but a sign of professional wisdom. Also, if you are threatened or harmed by a client, you should inform your field supervisor and faculty liaison immediately.

Of course, the best strategy for handling workplace violence is prevention. Talk with your field supervisor about issues of risk and safety specific to your internship site. Inquire about any policies or training that may be in place to protect worker safety.

 ◆ **EXERCISE 11.11 ANALYSIS**

What is your typical pattern of responding to people who are aggressive or threatening? Based on the information discussed in this chapter, how might you need to modify your typical response in order to intervene effectively with aggressive clients?

What risk factors for violent behavior, if any, exist within your agency and/or the populations it serves? What measures does your agency use to help ensure worker safety?

> **A Student's Reflections on Personal Safety**
>
> A child became very angry today and slammed a door really hard. A staff member was standing just inside the door with his hand inside the doorframe. It was obvious that the girl was trying to slam his hand in the door because we saw her glance at his hand right before she slammed the door. It was a heavy steel door and could have done some serious damage, especially with the force that was behind it when she slammed it shut. Fortunately, the staff member was really alert and quickly pulled his hand back. It was almost like he was reading her mind when she glanced at his hand in the door. If he had not been so alert and so attuned to her thinking, he could be missing some fingers right now or at least have some broken bones.

Conclusion

Even the most ideal students in the most ideal settings can have some difficulties occasionally in their internships. This chapter emphasizes that awareness of potential trouble spots can enable you to act wisely in avoiding them.

You can avoid many problems by observing good basic work habits, by maintaining healthy boundaries within your relationships, by striking a responsible balance between the demands of your work and those of your personal life, and by doing all that you can to keep yourself safe.

FOR YOUR E-PORTFOLIO

Write a letter to the next internship student who will be entering your field placement. Help this student recognize not only the opportunities for learning in this setting but also the potential challenges or trouble spots, and how to avoid them. As an attachment, include a tip sheet that you have created specific to your internship. Consider tips related to each of the major topics in this chapter including making mistakes, work habits, professional boundaries, dual relationships, balancing work and personal life, and safety. Feel free to include any additional categories of information that you feel would be useful based on your experience.

References

Alle-Corliss, L., & Alle-Corliss, R. (1998). *Human service agencies: An orientation to fieldwork.* Pacific Grove, CA: Brooks/Cole.

Alle-Corliss, L., & Alle-Corliss, R. (1999). *Advanced practice in human service agencies: Issues, trends, and treatment perspectives.* Belmont, CA: Wadsworth.

Barker, R. (1996). *The social work dictionary.* Washington, DC: NASW Press.

Bernard, J., & Goodyear, R. (1992). *Fundamentals of clinical supervision.* Boston: Allyn & Bacon.

Cherniss, C. (1980). *Staff burnout: Job stress in the human services.* Beverly Hills, CA: Sage.

Corey, M., & Corey, G. (2003). *Becoming a helper* (4th ed.). Belmont, CA: Brooks/Cole.

Faiver, C., Eisengart, S., & Colonna, R. (1995). *The counselor intern's handbook.* Pacific Grove, CA: Brooks/Cole.

Farber, B. (1983). Introduction: A critical perspective on burnout. In B. Farber (Ed.), *Stress and burnout in the human service professions* (pp. 1–22). New York: Pergamon Press.

Freudenberger, H. (1983). Burnout: Contemporary issues, trends and concerns. In B. Farber (Ed.), *Stress and burnout in the human service professions* (pp. 23–28). New York: Pergamon Press.

Griffin, W. (1995). Social worker and agency safety. In R. Edwards & J. Hopps (Eds.), *Encyclopedia of social work* (pp. 2293–2305). Washington, DC: NASW Press.

Hiratsuka, J. (1988, September). Attacks by clients threaten social workers. *NASW News,* 3.

Kaiser, T. (1997). *Supervisory relationships: Exploring the human element.* Pacific Grove, CA: Brooks/Cole.

Keith-Spiegel, P., & Koocher, G. (1985). *Ethics in psychology.* Hillsdale, NJ: Random House.

Kitchener, K. (1988). Dual role relationships: What makes them so problematic? *Journal of Counseling and Development, 67,* 217–221.

Littleton, N. (1996). Personal qualities in a successful human services career. In H. Harris & D. Maloney (Eds.), *Human services: Contemporary issues and trends.* Boston: Allyn & Bacon.

Murdach, A. (1993). Working with potentially assaultive clients. *Health and Social Work, 18,* 307–312.

Newhill, C. (1992). Assessing danger to others in clinical social work practice. *Social Service Review, 66,* 64–84.

Newhill, C. (1995). Client violence toward social workers: A practice and policy concern for the 1990s. *Social Work, 40,* 631–636.

Petrie, W., Lawson, E., & Hollender, M. (1982). Violence in geriatric patients. *Journal of the American Medical Association, 248,* 443–444.

Royse, D., Dhooper, S., & Rompf, E. (1996). *Field instruction: A guide for social work students* (2nd ed.). White Plains, NY: Longman.

Royse, D., Dhooper, S., & Rompf, E. (1999). *Field instruction: A guide for social work students* (3rd ed.). White Plains, NY: Longman.

Russo, J. (1993). *Serving and surviving as a human-service worker.* Prospect Heights, IL: Waveland Press.

Safety first is primary rule on job. (1998, November). *NASW News,* 4.

Schram, B., & Mandell, B. (2006). *An introduction to human services: Policy and practice* (6th ed.). New York: Allyn & Bacon.

Schulz, L. (1987). The social worker as a victim of violence. *Social Casework, 68,* 240–244.

Schulz, L. (1989). The victimization of social workers. *Journal of Independent Social Work, 3,* 51–63.

Shea, S. (1988). *Psychiatric interviewing: The art of understanding.* Philadelphia: Saunders.

Slimp, P., & Burian, B. (1994). Multiple role relationships during internship: Consequences and recommendations. *Professional Psychology: Research and Practice, 25,* 39–45.

Star, B. (1984). Patient violence/therapist safety. *Social Work, 29,* 225–230.

Sweitzer, H. (2004). Burnout: Avoiding the trap. In H. Harris & D. Maloney (Eds.), *Human services: Contemporary issues and trends* (3rd ed., pp. 339–353). Boston: Allyn & Bacon.

Sweitzer, H., & King, M. (2004). *The successful internship: Transformation and empowerment in experiential learning* (2nd ed.). Belmont, CA: Brooks/Cole.

Thomlison, B., Rogers, G., Collins, D., & Grinnell, R. (1996). *The social work practicum: An access guide* (2nd ed.). Itasca, IL: Peacock.

Truitt, M. (1991). *The supervisor's handbook: Techniques for getting results through others.* Shawnee Mission, KS: National Press Publications.

Vallianatos, C. (2001, June). Security training helps deflect assaults. *NASW News.* Retrieved November 26, 2005, from http://www.socialworkers.org/pubs/news/2001/06/security.htm.

Weinger, S. (2001). Security risk: Preventing client violence against social workers. Washington, DC: NASW Press.

Wilson, S. (1981). *Field instruction: Techniques for supervisors.* New York: The Free Press.

Chapter 12

Ending Your Internship

As you approach the end of your fieldwork, you might find leaving your internship to be every bit as challenging as beginning it was some time ago. Depending upon the nature of your internship site as well as the nature of your relationships with your supervisor, co-workers, and clients, leaving the internship may be an experience of great emotional intensity or one of relative ease (Sweitzer & King, 2004). In any case, the work of leaving a field placement of any type focuses on such tasks as ending your relationships in a positive way, reflecting upon and evaluating your experiences, examining your own growth and development, and writing any necessary closing and/or transfer summaries. In order to handle these tasks well, it is helpful to know as much as possible about the termination process. This chapter introduces you to important guiding principles for ending your fieldwork effectively and explores the specific tasks and processes that are necessary to bring your work to a positive close.

A Student's Reflections on Leaving the Internship

This has been the single most valuable experience I have had as a college student. I can't imagine anything that I could have done this semester that would have taught me more. I'm not ready to leave and I don't think the people at the group home (staff or clients) are ready for me to leave either. All of the children here have been left

continued

Ending Your Internship

271

again and again throughout their lives. I feel guilty about being another one of those people. They keep asking me why I have to go, when my last day is, what I'll be doing after I leave, and so on. I guess it helps them to talk about it, but I find it difficult every time it comes up. Today I was talking with a little boy whose parents have lost their parental rights. He is waiting to be adopted. As we were talking about my leaving he suddenly asked, "Will you be my mama?" This was extremely difficult for me to handle without getting upset, but somehow I muddled through it. I am finding that leaving this internship is the most exhausting part of the entire experience.

Evaluating Your Performance

Near the end of your internship, you and your field supervisor will engage in the important process of formally evaluating your work. Although you have been evaluated throughout your fieldwork and you have been evaluating and reflecting upon your own performance, these processes have been for the most part formative, that is, the evaluative feedback was for the purpose of educating you and improving your performance (Bogo & Vayda, 1995). The formal evaluation near or at the end of your internship is, in contrast, a summative evaluation. Although summative evaluations are also intended to be educational and can improve your performance, this is not their most immediate objective. Summative evaluations are primarily for the purpose of assessing overall outcomes and performance quality. Such evaluations look at the experience as a whole, identifying strengths and weaknesses (Alle-Corliss & Alle-Corliss, 1998). This chapter focuses on helping you to engage productively in this summative evaluation process.

A Student's Reflections on Evaluation

My supervisor had me fill out the evaluation on myself, and we compared it to his. We were basically similar on most things, but where we were different, I had evaluated myself lower than he had. He has given me feedback along the way, so the evaluation really wasn't too much new information for me. Overall, what I got out of the evaluation was a good feeling about myself and my work. He was very positive about most things, but I also know the things I need to continue working on. I am pleased to say that I can look at my performance with pride.

Understanding the Final Evaluation

To some extent, evaluation has occurred throughout your internship as part of your ongoing supervision and learning process. Therefore, there are usually few surprises for students at the time of final evaluations. This is not always the case, however, due to a number of factors. The thorough and comprehensive nature of the final evaluation process brings the entire internship experience under review. In doing so, patterns, themes, and issues might emerge that were less obvious in the day-to-day work with your supervisor. Also, you might find that although you and your supervisor have discussed your strengths and weaknesses openly over the term, the evaluation calls upon both of you to weigh and balance these areas in relation to one another. In this process, you might find that what one party perceives as a minor issue, the other party may see as quite a significant one, and vice versa. Students and supervisors sometimes find that they weigh the importance of various factors differently and perceive the balance between strengths and weaknesses differently.

An important issue in understanding the evaluation process is accountability. Accountability has to do with the responsibilities of the various parties involved in the evaluation and might be defined as the "state of being answerable to the community, to consumers . . . , or to supervisory groups . . ." or as the "obligation of a profession . . . to provide assurances that its practitioners meet specific standards of competence" (Barker, 1996, p. 3). Kaiser defines accountability as "the process of taking responsibility for one's behavior and for the impact of that behavior on self and others" (1997, p. 14). She says further that accountability involves "a commitment to tell the truth" and "a commitment to take responsible action" (Kaiser, 1997, p. 14). In recent years, concerns about issues of accountability in human service agencies as well as in educational institutions have increased. Human service educators and field supervisors bear the responsibility for ensuring that those individuals entering the human service field meet certain minimum standards of competence (Simon, 1999). Therefore, student evaluations, as well as evaluations of practitioners in the field, must be objective and adhere rigorously to certain standards. All parties involved in your final evaluation—you, your supervisor, and your faculty liaison—must understand these issues of accountability in order to ensure that your evaluation is as objective and as fair as possible.

Evaluation is almost always an anxiety-provoking experience, regardless of the quality of your work (Bogo & Vayda, 1995; Wilson, 1981). It might be helpful to recognize that this type of evaluation is not unique to your being a student. All human service professionals are evaluated regularly by their supervisors, usually on an annual basis at minimum. Therefore, developing the ability to engage productively in the evaluation process should be seen as part of building your repertoire of professional skills. As you approach your evaluation, your expectations about it and reactions to it will tend to be shaped by your previous experiences with evaluation. The following exercise will help you to identify the expectations that you bring into the evaluation process.

◆ EXERCISE 12.1 PERSONAL REFLECTION: OBSERVATION OF SELF AND OTHERS

Think of a few instances in the past in which your work has been evaluated. You might draw upon experiences within jobs you have held or within previous fieldwork.

To what extent did you experience these evaluations as fair and objective? What are your expectations of your evaluation in this field experience based upon your previous experiences with being evaluated?

As you anticipate your evaluation in this field experience, what aspects of the evaluation process do you feel most anxious about? Most confident about?

As a developing professional, you are expected to acquire the ability to evaluate your own performance objectively and participate in your evaluation. In previous evaluations, how actively did you participate in the evaluation process? How comfortable are you with your ability to participate actively in this evaluation?

Preparing for Your Evaluation

As evaluation time approaches, you should prepare yourself to be an active participant in your evaluation by reflecting upon the quality of your performance (Thomlison, Rogers, Collins, & Grinnell, 1996). The first step of this process will be assembling the various tools that will be used in the assessment process. These tools include

◆ the learning plan or contract that you developed at the beginning of your internship,
◆ the course syllabus or other document that lists the course objectives,

- the student evaluation instrument used by your human service program,
- any notes, documentation, tapes, or other concrete materials that reflect your work during the internship, and
- your calendar or day planner for the time span of your internship.

Your learning contract, the course objectives, and the student evaluation instrument function as a collective set of criteria for evaluating your performance (Simon, 1999). To evaluate your performance, focus on each goal or outcome identified in these documents. Try to evaluate your own behavior in relation to each criterion. Referring to your notes, tapes, calendar, and other materials helps you to respond objectively to each item, identifying supporting information for whatever rating you might assign yourself.

Without such objective consideration of your own performance and behavior, it is easy to lapse into some common rating errors. A rating error is "the tendency to make a global judgment about a worker's performance and then to perceive all aspects of the worker's performance as consistent with that general judgment" (Kadushin, 1976, p. 287). One of the most common rating errors is the halo effect. A halo effect occurs when students feel that they have done a good job in their field placement and therefore assume that each and every item on their evaluation will be rated positively, with any negative feedback seen as unfair. Although a student's global self-assessment as "good" might be accurate, it is an obstacle if this generalization interferes with a closer, more objective look at the relative strengths and weaknesses of the student's performance on particular skills. Looking at objective data regarding your work during the internship can help reduce the tendency toward rating errors and enable you to see your own performance more clearly.

◆ EXERCISE 12.2 ANALYSIS

Review the learning contract that you developed for your internship, the course goals and objectives for the field experience, and the student evaluation instrument that your field supervisor will use to evaluate your performance. Rate your own performance using these various criteria.

Drawing upon this information, write an open-ended evaluation of your performance in your internship, organizing your thoughts into the following three paragraphs or topics: (1) your greatest strengths in this field placement, (2) your greatest weaknesses during this field placement, and (3) your goals for professional growth and development when you enter your next field placement or job.

When you have completed Exercise 12.2 thoroughly and thoughtfully, you should be well prepared for your field supervisor's evaluation of you and ready to discuss your evaluation in an objective manner. The final steps in preparing for your final evaluation are (1) considering the roles, perspectives, and obligations of your field supervisor in the final evaluation process, (2) developing an understanding of your evaluation conference with your supervisor, and (3) considering the roles, perspectives, and obligations of your faculty liaison in participating in your evaluation and assigning your final grade.

The Field Supervisor's Perspective

Like students, supervisors also often find evaluation to be an anxiety-provoking process. Field supervisors who have developed a level of trust and closeness with a student may find it difficult to get enough distance to evaluate that student objectively.

They may want to avoid potential conflict with the student by avoiding difficult issues or less glowing aspects of the student's performance. The evaluation process might also prompt self-evaluation in the supervisor. Supervisors might begin to question their own methods of working with students and wonder whether their student might have been more successful if they had been supervised differently.

Supervisors also bring to the evaluation process their own experiences with evaluation. The evaluation standards, processes, and methods used within the settings in which the supervisor has worked undoubtedly cast a larger context in which student evaluation is viewed. For example, a supervisor who is working or has worked in settings in which worker evaluations were rigorously conducted is much more likely to conduct rigorous evaluations of students. In other words, field supervisors bring their own humanness and experiences to the evaluation process and often have their own thoughts, feelings, and histories to sort out in relation to the evaluation process. Although every student might hope for a supervisor who simply pats them on the back, writes a consistently glowing evaluation, and sends them happily on their way, most students also realize that this type of evaluation in the end is not very helpful.

The purpose of your evaluation is to help you leave your placement with a clear grasp of your current level of skill and knowledge as well as with some clear goals for your future learning, growth, and skill development. In order to achieve this purpose, your supervisor must rely upon objective, measurable, observable data about your work in completing your evaluation (Faiver, Eisengart, & Colonna, 1995). An evaluation that is too vague or general will not give you the feedback that you need either to recognize your own strengths or to improve your future performance. Neither does an evaluation that is too one-sided, addressing only strengths or only weaknesses, give you this focus. With all of this in mind, expect that you and your field supervisor will identify areas of strength as well as areas that need to be improved and that these impressions will be drawn from objective observations of your work. Given that this is the expected range of focus for any evaluation, suggestions for future improvement should not be taken as criticism. Even the most experienced and proficient practitioners can identify areas of their performance that need to be improved.

To understand your supervisor's perspective, it is helpful for you to keep in mind the issues of accountability affecting your supervisor's evaluation of you. As discussed earlier, field supervisors have responsibilities not only to their students but also to other constituencies as well. Their evaluations must give consideration to the future clients and organizations that the student will serve as well as to the integrity of the profession as a whole. Also, your field supervisor is accountable to your academic department. An important duty of the field supervisor is to give the human services faculty objective, honest feedback about each student's abilities and learning needs. As your supervisor completes your evaluation, all of these levels of responsibility come into play.

The Evaluation Conference

Most field supervisors complete the written evaluation and share it with their students in some form. For your evaluation to have the greatest learning benefits for you, you should have the opportunity to discuss it with your supervisor. In light of the earlier discussion of how difficult evaluation can be for both the student and the supervisor, it is perhaps understandable that at times one or both parties may try to avoid this direct discussion. If your supervisor does not offer to discuss your evaluation with you, it is appropriate and advisable for you to request a discussion (Chiaferi & Griffin, 1997). Useful discussion topics for this conference include any particular questions you might have about your supervisor's ratings or assessments, areas in which your perceptions and your supervisor's perceptions differ, your and your supervisor's perceptions of your greatest strengths and weaknesses, and ideas about how you might improve your performance in your next experience, whether this should be another educational fieldwork experience or a job.

A major goal for you as the student in this conference is to develop a clearer and more detailed understanding of how your work has been perceived by your supervisor. Through this discussion, you can possibly develop a fuller and more complete picture of your performance, one that is enhanced by seeing yourself through the eyes of a more experienced professional worker. A further goal of the conference is to develop some ideas regarding how you might use your supervisor's feedback to focus and direct your future growth and development. Having said this, it should be clear that the discussion is not for the purpose of trying to modify your supervisor's evaluation or perceptions, although this sometimes occurs as a by-product of the discussion. It is appropriate for you to point out any concrete evidence you might have to support a rating different from your supervisor's (Thomlison et al., 1996).

Although the conference needs to be a conversation involving two-way communication between you and your supervisor, keep in mind that the conference will have the greatest value for you if you can enter it with a spirit of openness and flexibility. You will want to share your own perceptions honestly and straightforwardly with your supervisor, but it is also important to be ready to listen, recognizing that the conference is potentially one of the most important learning experiences of your internship. Therefore, it is wise to enter the conference with a nondefensive posture, with your mind and ears open, so that you can hear what your supervisor has to say.

A Student's Reflections on the Evaluation Conference

I was a little nervous going into my evaluation conference. I didn't think there would be any big problems, but still evaluation time is kind of like test time and it always makes me nervous. I was relieved that the conference went great! She reassured me about my goals and my career choice. Even though I know that this is the field I want, it is wonderful to hear an experienced professional say that I will be good at it. She noted particularly my strengths in working with children, my writing skills, and my ability to get work done promptly. We also discussed the fact that I need to continue developing my computer skills and my assertiveness in certain situations. Overall, I felt it was a very positive evaluation but one that did not gloss over the "needs to improve" category. I appreciated her thoroughness and honesty in discussing the evaluation with me.

◆ EXERCISE 12.3 PERSONAL REFLECTION: OBSERVATION OF SELF AND OTHERS

Complete this exercise after you have had your evaluation conference with your supervisor. In reflecting upon this discussion and the feedback you received, discuss the following questions: What messages were you most pleased to receive from your supervisor? What messages were most difficult for you to hear or accept? How well were you able to maintain an open and nondefensive posture throughout the discussion?

How might your supervisor's feedback clarify or reshape your goals for future improvement and growth?

If you were to have a job interview today, what particular skills and strengths could you confidently tell a potential employer that you demonstrated during your internship?

The Faculty Liaison's Perspective

The faculty member from your academic department who works with you in your internship might be referred to by a number of titles, including the faculty liaison, faculty supervisor, or internship instructor. Whatever the title, this individual has an important role in determining your final grade for your fieldwork. The faculty member, as a college or university employee, is responsible for assigning the final grade, taking into consideration the field supervisor's evaluation as well as your performance on other requirements of the course, such as written assignments. How various components of your performance are weighted varies from program to program and may even vary some from one faculty member to another.

As discussed earlier, important issues of accountability come into play as your field liaison determines your final grade. Your faculty liaison is responsible for upholding the standards of the academic department and is acutely aware that every student in the field represents both the college and the academic department. In fact, the faculty member at some point in the future probably hopes to work with your field agency and your field supervisor again as they host another student intern in the future. Each student's performance in the internship can pave the way for the next student, make the next placement rockier, or even make another placement in the agency impossible. Clearly, the academic program has much to gain as a result of those students who perform well in the field and much to lose as a result of those who do not.

The credibility and integrity of the academic department rests in part upon the program's ability to attest to the quality of its graduates. Because field liaisons work with many students in the field, they are perhaps in the best position to weigh the quality of students' performance in relation to one another, identifying those students who have done an outstanding job as compared to those whose work was average, minimally adequate, or unsatisfactory. In subsequent months and years, the student's grade in the internship may be included among other materials as various students are considered for graduate school admission or for employment. The faculty member's personal credibility, as well as that of the department, are damaged if the grade does not honestly and accurately reflect the student's ability and the quality of the student's performance.

As a human service professional, faculty liaisons (like field supervisors) are also accountable to the profession. They have a stake in ensuring that all those entering the field perform adequately to meet the needs of clients as well as the expectations of the employing agencies. From an ethical perspective, the faculty member might be seen as a gatekeeper of the profession, ensuring that only competent practitioners earn the credentials of the profession and recognizing that incompetent practitioners damage the profession as well as the clients they serve and the agencies that employ them.

Finally, your faculty liaison is accountable to you. Your faculty liaison wants you to be successful in your internship and no doubt would like you and every student from the program to *earn* the highest possible evaluation. You can be sure that this faculty member wants to do everything possible to support your success throughout the internship. As a faculty member in your academic department, she or he may have a longstanding relationship with you. This relationship might make objective and rigorous evaluation of your performance particularly difficult. Nevertheless, faculty liaisons' accountability to students in the evaluation process requires that they be honest and objective in giving students feedback. Your faculty liaison and field supervisor share responsibility for ensuring that you leave your fieldwork with an honest and realistic picture of your strengths and weaknesses as a practitioner, especially since this information should guide your continuing professional development. The responsibilities of both the teaching profession and the human service profession require the faculty member and the field supervisor to formulate honest, objective assessments of your performance and your level of competence as a developing human services professional.

A summative evaluation of your work is an important component in ending your internship productively. This evaluation process calls upon you to take an active part in assessing your learning and your performance. You, your field supervisor, and your faculty liaison have important roles to play in your final evaluation. Similar evaluations will continue throughout your career because evaluation is an essential component of accountability within the profession and your own professional growth and development.

Leaving Your Internship

In addition to evaluation processes, you will be taking part in other processes that surround the ending of the internship experience. Just as with the evaluation process, ending your internship should be done with a high degree of self-awareness, intentionality, and professionalism. Ending relationships can be difficult. The process of leaving your internship involves ending not just one relationship, but many. As a result, this period can challenge your interpersonal skills and your emotions. The technical term generally used in discussing such endings in human services contexts is *termination*. Termination is "a systematic procedure for disengaging the working relationship" (Barker, 1996, p. 380). The termination stage of the internship was briefly discussed in Chapter 1, Getting Ready, as you were introduced to the stages of internship development. Just as the end of the internship was discussed at the beginning of this book, so your termination from the internship has been anticipated from its beginning. You, your clients, your supervisor, and your co-workers have known since the outset that your stay in the organization was time-limited. Possibly there was from the beginning a very specific predesignated date for ending your fieldwork. In spite of this, the time to leave your internship can seem to come upon you suddenly and before you are ready for it.

Often, students express resistance to ending with comments such as, "I was just beginning to get good at this. I don't want to leave now!" At the same time, students often express great pleasure, pride, and satisfaction in ending their fieldwork due to the rich and valuable experiences that they will take with them and their sense of having performed a job well. Feelings of accomplishment and pride are often mingled with feelings of sadness, regret, and even relief as the experience draws to a close. Such reflections and mixed feelings are predictable and normal aspects of the termination process.

Conducting a positive termination process calls upon you to engage in extensive reflection on many aspects of your experiences over the course of the internship. Your formal performance evaluation is one aspect of this reflection. You will probably be asked to evaluate your field placement and your field supervisor as well. In preparing to leave your agency, you also need to think carefully about each client and each project that you have been working with. Plans need to be made for work that requires continued attention after your departure. Throughout the termination process, you will also be involved in the sometimes difficult emotional work of saying good-bye, allowing yourself and others to feel and express whatever emotions are present as you go about the business of leaving. Understanding the principles involved in termination enables you to leave your internship in a manner that enhances your learning and brings closure to your work.

General Guidelines for Positive Termination

In your academic coursework, you have probably studied how to terminate with clients effectively. Reviewing any coursework you have done on this topic or seeking out relevant reading is a good idea as you begin terminating your internship. Many of the principles that guide termination of the helping relationship are helpful to you here. The following principles of positive termination of your internship are adapted from guidelines for handling termination within helping relationships. You are more likely to have a positive termination of the internship if you follow these principles not only in ending your relationships with your clients but also in working through your own termination from the internship.

Be Aware of Your Previous Experiences and Patterns with Terminations

All of us have dealt with losses. In the course of normal life, there are many relationships and situations that ultimately end in separation and saying good-bye. Some of these events are intense and difficult, whereas others are easier, perhaps just a normal part of growing up. Moving away from friends and family, changing schools, breaking up with a romantic partner, and experiencing a divorce or the death of a loved one are all common life events that people frequently have to deal with. Over the course of dealing with such separations, individuals tend to develop patterns in their reactions, feelings, and behaviors surrounding separations.

Identifying and evaluating your own patterns in dealing with endings is a good first step in preparing to handle the many terminations inherent in leaving your internship. Likewise, it is helpful during this time to engage clients in similar exploration, helping them to identify their own patterns in saying good-bye. Clients who can approach termination with this degree of self-awareness can often feel more in control simply because they can better anticipate the feelings and reactions that they are likely to experience along the way.

A Student's Reflections on Termination Patterns

In my personal life I generally try to avoid good-byes. Whenever I've moved or left friends behind, I always kind of hide behind a "see you later" type attitude. I make lots of promises to keep in touch and the other people do the same, but it usually doesn't happen. I will be graduating from college soon, and already I am saying these same lame lines to my friends. Of course, true to form, this is also my tendency in my internship as well. It is really tempting to say I'll keep in touch, come by to visit, and so on. I realize that this is where I have to draw the line with my old patterns. I can't make easy, false promises here. I need to be honest and say a real good-bye. I hate doing that because it is so sad.

 ◆ **EXERCISE 12.4 SYNTHESIS: LINKING KNOWLEDGE AND EXPERIENCE**

Identify a few particularly meaningful separations that you have experienced previously. What were your emotional reactions to these separations? What were your behavioral reactions? What strategies did you use to deal with the losses involved in these separations?

How might these patterns be helpful or unhelpful as you terminate your internship? In what ways might you need to alter these patterns in order to terminate from your field placement in a positive and healthy manner?

Be Self-Aware, Recognizing Your Needs and Wants

(Baird, 1999; Patterson & Welfel, 2000; Sweitzer & King, 2004)

A range of reactions is normal at the time of termination. Being aware of these reactions is a necessary step in managing terminations effectively. In terminating with clients, they need to be encouraged to recognize and express their wants and needs as they anticipate the end of the relationship. These wants and needs may not be fixed and static, because clients often express different reactions at various points in the termination process or even simultaneously. At times, their feelings might even be contradictory and conflicting. Such ambivalence (a state of mixed feelings and reactions) is a normal response to termination (Chiaferi & Griffin, 1997). Because ambivalence can create confusion and distress, it is especially important to explore such feelings and offer emotional support.

Just as clients benefit from addressing their needs and wants at the time of termination, you can do likewise. At times, you might be eager to terminate, ready to move on to the next challenge in your life or ready to take a break. At other times, you may feel resistant to leaving your placement. You might feel that you have not yet learned enough, or you might regret that you have not been as involved as you could have been. Being aware of your wants and needs can help you to avoid some common pitfalls in termination, such as premature disengagement from your internship (discussed in Chapter 1) or denial of the pending termination. Students who disengage prematurely from the internship tend to withdraw from clients and co-workers and may even become less attentive to their work responsibilities as the end of their internship approaches. Students who deny the pending termination may fail to give clients adequate warning of their leaving or may give clients false reassurance by promising that they will keep in touch. Therefore, it is important to be honest with yourself about your wants and needs regarding leaving your internship so that you can avoid such pitfalls in your work. During termination, students, like clients, can also feel confused and ambivalent about what they want and need.

A Student's Reflections on Wants and Needs

At times I find myself desperately wanting a job here. At these times I simply don't want to leave. The fact that there is some slim chance that I might actually get a job here within the next several months makes it all the more difficult to let go. At other times I am ready to go. I think the main reason I feel this way at times is that I am just very tired. I need a break. I am very grateful for the opportunity I have been given here, but I do feel like it is time to stop being a student. As much as I've liked my internship, it is a little bit like "Neverland." I'm not a kid and not a grown-up, not an employee and not a volunteer. I want to be a full-fledged, employed, paid professional somewhere, and most of the time I would like that place to be here.

Reflect Upon and Deal with Your Feelings

(Alle-Corliss & Alle-Corliss, 1998; Brill, 1995; Okun, 2002; Sweitzer & King, 2004)

Clients need to be encouraged to express their emotions about the approaching end of the helping relationship. Emotions around termination can cover a wide range from sadness, anger, resentment, and frustration to relief, joy, satisfaction, and pride. For both clients and you, some of these feelings may be easier to acknowledge and discuss than others. Give yourself and others the permission to experience the full range of emotions that may be present.

Whatever your feelings are, it is helpful to verbalize them to someone who is a good, accepting listener. Your field supervisor may want to explore your feelings about termination with you. Also, discussing your feelings with other internship students is often very helpful because you are all going through similar experiences. The field seminar is an ideal opportunity for this type of discussion. Discussing your feelings is an effective way of becoming more comfortable with them and bringing emotional closure to your internship.

♦ **EXERCISE 12.5 PERSONAL REFLECTION: OBSERVATION OF SELF AND OTHERS**

As you approach the end of your internship, what are your wants and needs? In what ways are you ready to end the internship? In what ways are you resistant to leaving?

What emotions are you experiencing as you anticipate the end of your internship? Which of these emotions are you most comfortable with? What emotions, if any, are you uncomfortable acknowledging or discussing?

Review the Experience
(Alle-Corliss & Alle-Corliss, 1998; Okun, 2002; Young, 1998)

As clients engage in the termination process, they are invited to review the helping process, identifying the critical events that led to positive change for them. These events are sometimes thought of as turning points in the client's growth and development. It is not uncommon for clients at the end of the helping relationship to look back at some of the most difficult or discouraging points of the process as being most important in retrospect.

As you have experienced your internship day-to-day, you have probably been aware of events that seemed particularly exciting, interesting, or educational. At other times, you might have found particular events to be tedious, boring, stressful, or even distasteful. In the immediacy of the everyday demands, it can be difficult to recognize the value of particular events. Toward the end of your placement, as you reflect on your experience as a whole, you are in a better position to identify its most valuable components. Surprisingly, some of the most valuable events of the internship might be those that at the time were unpleasant or seemed to be of little consequence. Conversely, those events that seemed very significant or positive at the time might over the long run prove not to be so important to your overall development and learning.

A Student's Review of Key Events

It is really ironic that I chose this internship because it dealt with juvenile crime. In fact, this was about the only thing I even knew about the placement before I started it. As it turns out, this has probably been its least important feature. What I have gotten interested in and learned about here has been family dynamics and parenting issues. The problem focus could have been just about anything. What I have gotten out of this has been so much broader and deeper than just learning about a single population or problem. It has opened my eyes to a whole new way of understanding human behavior. My goals have changed dramatically as a result. I am exploring graduate programs in family counseling now.

Acknowledge the Progress and the Changes That You Have Made
(Baird, 1999; Cormier & Hackney, 1993; Royse, Dhooper, & Rompf, 1999)

When clients terminate from a helping relationship, it is especially helpful to reflect with them upon the gains they have made during the course of the helping relationship. Acknowledging these gains, being able to name them and claim them, helps clients to retain their progress once the relationship has ended. Furthermore, recognizing their gains enables clients to leave the helping relationship with a sense of pride and accomplishment. A similar process will be helpful to you as you leave your internship. It is important that you think about your development over the course of the internship, focusing particularly on what you have learned and how you have grown and changed, professionally and personally. As with clients, recognizing and acknowledging the progress that you have made in your own growth and development gives you a firmer grasp of what you have learned and a sense of pride and accomplishment.

A Student's Reflections on Accomplishments

I divide what I've actually achieved in this internship into two categories: what I achieved for myself and what I achieved for the agency. For myself, I feel that I have made the big shift into professional life. I've learned to hold my own in a group of professionals, to carry my own workload competently, to handle difficult people, and to juggle many tasks. I have learned how to work on multiple projects, with multiple teams, and multiple bosses. I know that I can go into a job interview now and talk about my abilities in a more convincing way because now I really know that they are there. For the agency, I have organized a county-wide agency fair (almost single-handedly), written a small grant for some playground equipment, and led a major fund-raising event. In addition, I hope and believe that I have offered some support and encouragement to a number of children who really needed it. Have I changed the world? No. Have I made a difference? I can say with confidence, "Yes!"

To fully appreciate what you have accomplished during your internship, it might be helpful to think about your first days there and compare them to your experience now. In doing so, it might also be helpful to review some of your writing in response to earlier exercises in this book. Such a review probably reveals some fairly significant learning and growth over the course of your internship.

 ◆ **EXERCISE 12.6 ANALYSIS**

A key question to consider as you reflect upon your internship is, "What particular events in or components of this internship have caused me to learn, grow, and change?" Identify at least three specific events in your internship that have been particularly meaningful in your experience. In the case of each event, try to articulate what gives the event special significance.

As you look at your development over the course of the internship, in what specific ways have you grown, learned, and changed during the internship?

Imagine that you are interviewing for a human services job or for your next field placement and that the interviewer asks, "So what did you get out of that internship?" How would you respond to such a question?

These guiding principles for termination emphasize the importance of intensive review, reflection, and evaluation as you leave your field placement. Keep these principles in mind, remembering to apply them to your clients as well as to yourself, as you go about the tasks involved in termination. They will be particularly helpful to you as you bring the most meaningful relationships of your internship to a close.

Saying Good-Bye to Your Supervisor

Although you might have worked with many professionals during your internship, your field supervisor probably holds special significance in your experience. This individual, more than any other, generally has the greatest impact on student experiences in the field. It has been argued, in fact, that the student–supervisor relationship is of primary importance in student learning, offering the best medium for preparing students to meet the demands of practice situations (Matorin, Monaco, & Kerson, 1994).

Because of its significance, your relationship with your supervisor should be terminated in a manner that reflects the general principles for positive termination. Many supervisors ensure that this occurs by initiating discussions in which the high points and low points of the supervisory relationship are explored, the student's growth and development within that relationship are evaluated, and feelings about ending the internship are expressed. Such discussions with your supervisor are beneficial for you as an intern in two ways. First, these discussions provide a structured time and opportunity in which you can explore and work through the process of termination. Second, such discussions offer you the opportunity to see your supervisor modeling the steps and professional skills that are involved in effective termination (Bogo & Vayda, 1995).

As you and your supervisor reflect on your internship experience, issues related to the evaluation of your supervisor and of the agency as a placement can sometimes surface (Sweitzer & King, 2004). Your academic program probably requires you to complete a formal evaluation of your field placement and/or your supervisor, which usually involves completing an evaluation form so that systematic information can be gathered by your program about all of the various field sites that host student internships. In addition, your field supervisor might ask you for feedback about the placement. Many field supervisors want to know how they could improve the experience as well as what components of the experience were particularly valuable. Although not all supervisors ask for such feedback, those who do generally are quite sincere in wanting your honest assessment. This can, however, be an awkward experience for you as a student. If you should be in this position, try to remember that just as you cannot learn from your evaluation unless it is shared and discussed with you, field supervisors cannot benefit from feedback that is not shared with them.

Understandably, giving your supervisor feedback might be difficult for you to do in certain circumstances. If you feel, for example, that your placement did not fully meet your learning needs due to the nature of the agency's mission, this can hardly be attributed to your supervisor's handling of your internship. Therefore, such feedback will probably not be too difficult for you to share. If, however, you feel that your supervisor withheld certain types of experiences from you that were potentially available, this might be more difficult to address. Nevertheless, such issues are important to discuss.

In such a discussion, your supervisor might be able to point out particular aspects of your academic preparation or internship performance that suggested you were not ready for these experiences. This feedback could be enormously helpful to you as you prepare for your next field placement or for employment. By the same token, your feedback might prompt your supervisor to reevaluate the duties assigned to you and possibly modify them for the next student who enters the agency. In either case, the conversation will have yielded important benefits. Although it may be difficult to acknowledge the less positive aspects of your placement, it is beneficial to discuss them so that both you and your supervisor can learn from these events.

Because you are the student, more accustomed to getting feedback from your supervisor than giving it, you will probably feel the need to think carefully as you prepare for this conversation, identifying clearly the issues you would like to address and considering how to express yourself tactfully. Just as your expectation was that your supervisor would recognize your strengths as well as your shortcomings as a student, you would do well to strive toward the same balance and fairness of mind in your comments. Some field supervisors and placements, like some students, are of such high quality that it is indeed difficult to make suggestions for improvement. Even so, students in such placements should try to identify some ways in which the supervisor and the agency could further improve the quality of their work with students.

 ◆ **EXERCISE 12.7 PERSONAL REFLECTION: OBSERVATION OF SELF AND OTHERS**

Identify at least two of your supervisor's most important strengths in his or her role as an internship supervisor. To identify these, you might think of the qualities that you particularly appreciated about your supervisor, qualities from which you benefited as a student. Also identify at least two specific suggestions as to how your supervisor might have further enhanced your learning experience.

In a similar fashion, identify at least two particular strengths of the agency as a field site for students. Also make at least two suggestions as to how the agency could improve its work with students in the future.

Saying Good-Bye to Your Clients

Most field placements involve at least some direct work with clients. Whether these clients have been children in a child care center, residents in a group home, older adults in a senior center, inmates in a prison, employees in a company, or students in a school, you have come to mean something to those with whom you have worked, and they have come to mean something to you. If your field placement has been more administrative in nature, you have perhaps had less contact with traditional client groups and have instead worked more closely with staff or served other populations. These relationships, however, may be equally meaningful and difficult to leave behind. Therefore, all concerned, those with whom you have worked as well as you yourself, need an adequate opportunity to prepare for your leaving.

For many students, one of the most difficult aspects of leaving the internship is terminating with clients. The process is much smoother, of course, if you have been clear with your clients from the outset that you were a student in a time-limited placement. Clients need sufficient warning that the end of your relationship is approaching and a clear date as to when your last contact will occur. It is often helpful to taper off your relationship by gradually reducing the frequency and/or the length of your contacts. Such a process can help to reduce client anxiety about termination (Okun, 2002). For some clients, such a transition might not be warranted. These clients might continue working with you until the end of your internship, then transfer to another worker at that time. In this case, the client still needs a chance to terminate with you and to adjust to the idea of a new worker. In any case, you need to introduce the idea of termination early enough to make whatever transition is necessary in the client's care and to ensure that the client does not feel that your relationship was abruptly cut off.

In some settings, such as residential programs and group treatment settings, you might have worked as part of a treatment team without responsibility for ongoing, one-to-one contacts with clients. In such cases, there will probably not be a designated worker to replace you. Clients in such settings are not transferred to new workers and benefit from the continuity of the other team members' ongoing availability. Nevertheless, worker–client relationships within such settings can become quite close and meaningful. Consequently, your clients need the opportunity to engage in the termination process with you as you leave your internship. Upon your departure, remaining team members might be asked to provide additional informal support for those clients whom you anticipate having the most difficulty with your leaving.

A Student's Reflections on Saying Good-Bye to Clients

Since my internship has been in a senior center, I honestly didn't think my leaving would be a big deal. A lot of the termination stuff we have studied I thought was mostly relevant to casework or counseling situations. Much to my surprise, there have been lots of different reactions to my leaving. One of the ladies got teary-eyed a few days ago when we were talking about my last day. Mr. T., whom I have been playing cards with almost every day, seems to be avoiding me now. They all tease me about leaving, saying things like, "Go ahead and just abandon all us old folks!" I know they are teasing, but I still feel bad when they say it. I also know, as my mother always said, that "many a truth is said in jest." I'm going to talk to my supervisor about having a party or something before I leave. I think we need to do something to make it easier.

 ◆ **EXERCISE 12.8 ANALYSIS**

How well prepared are your clients for the end of your internship? To what extent have they been aware throughout your internship that you would be leaving at a specific point in time? Are there particular individuals for whom you feel your leaving might be especially difficult? If so, how might you make a special effort to prepare them at this point?

In terminating with clients, there are several tasks to accomplish. All of them flow from the guiding principles for termination discussed earlier in the chapter. As suggested there, you should engage clients in reflecting upon their work or relationship with you. This reflection should include evaluating their progress and identifying key events that enabled them to make progress. In addition, clients should be encouraged to discuss their feelings about termination and to make any plans for the future that are needed to support their continued stability and growth.

A complicating factor in terminating with clients is the fact that each individual has her or his own history of losses and patterns in dealing with those losses. As a result, you might find that no two clients react in exactly the same way to your leaving. Just as you explored your own history and patterns in saying good-bye earlier in the chapter, it is useful to explore possible reactions from clients as well. Although some clients may have intense emotional reactions to termination, most evidence suggests that client reactions are generally positive. Clients are much more likely to have positive reactions to termination if the ending has been anticipated and planned for systematically (Fortune, 1995). Clearly, the worker can do a great deal to ensure that termination is handled in this manner. Nevertheless, Barker says that terminating with clients often prompts emotional upheaval and cites client resistance, denial, and flight into illness as common client reactions (1996, p. 380). Each of these responses will be briefly considered.

Resistance occurs when clients try to avoid or impede the worker's efforts to discuss termination. Clients might resist, for example, by forgetting that you are leaving or forgetting the date of your departure. They might change the subject when the topic of termination is brought up. Likewise, the resistant client might fail to show up for appointments, particularly your last contact or the appointment in which the new worker is to be introduced. In such cases, it might be useful to point out to the client the pattern that you see and to discuss his or her patterns of saying good-bye in previous situations. Empathizing with how difficult it can be to let go and say good-bye can help give the client the emotional support needed to confront the situation. Resistance may be particularly intense in relationships that have been especially close and important to the client.

In contrast, some terminations may be marked by a lack of emotional intensity. This might be due to a lack of intensity in the relationship or due to client denial. For example, a minimal reaction could be predicted and is understandable from clients with whom you have had little contact or those with whom your work has been fairly limited. Many helping relationships, however, are characterized by personal disclosure and trust that have developed over a significant period of time. In such relationships, more emotional reactions from clients are to be expected. Therefore, it is significant when in such a context a client insists upon having no feelings about ending the relationship. Such clients may be denying their genuine feelings as a defense against potential emotional pain. In such a situation, it might be useful for you to express your own feelings of sadness and other genuine feelings about saying good-bye. Of course, it is important to keep this self-disclosure within appropriate limits. Keep in mind that your comments are designed to be helpful to the client and are not for the purpose of venting your own feelings or gaining emotional support for yourself (Cournoyer, 2005). Also, discussing with the client the range of emotions that clients often experience about termination can help normalize those feelings, possibly decreasing defensiveness and increasing the likelihood of more open discussion.

Another possible reaction to pending termination is sometimes referred to as *flight into illness*. This phenomenon occurs when the client suddenly begins to introduce new problems, relapse into old problems, or report a decrease in coping ability as the helping relationship is coming to an end. This reaction seems to be an effort to reengage the worker in the helping process and may be a sign of overdependence in the relationship or a fear of abandonment on the part of the client (Barker, 1996). In some cases, it seems likely that the client's increased anxiety in the face of termination might in fact create some temporary "new problems" or reduced coping ability. In any case, discussing such relapses with clients as a possible reaction to the pending termination can often help to allay their fears as well as help them to better understand their own feelings and behavior.

Knowledge of the three patterns discussed above—resistance, denial, and flight into illness—will help you to understand some of the varied client reactions that you might encounter toward the end of your fieldwork. Such reactions may present special challenges in your efforts to terminate with particular clients.

Many terminations, of course, are fairly straightforward because clients directly express appropriate and predictable feelings regarding termination. Even these more straightforward situations, however, can be challenging for you as clients' emotional reactions can trigger your own feelings of sadness, guilt, and regret. Common client reactions to termination include feelings of sadness, anger, loss, abandonment, and/or betrayal. Those clients who have had multiple, difficult losses might find termination particularly painful and emotionally intense, even if they have not had a very close relationship with you. Such a reaction is sometimes referred to as *transference*, because the client seems to transfer feelings from previous relationships onto the relationship with the helper. Whatever the nature and source of the client's feelings about termination, your role is to convey your empathy and acceptance of those feelings and to encourage the client to express and explore whatever feelings are present. This process of emotional expression is an important part of bringing your relationships with clients to a positive end.

 ◆ **EXERCISE 12.9 PERSONAL REFLECTION: OBSERVATION OF SELF AND OTHERS**

What emotions and reactions have you observed among the clients you serve as your termination approaches? How have these feelings been conveyed verbally and nonverbally?

In what instances, if any, have you been able to observe signs of denial, resistance, flight into illness, and/or any other problematic patterns?

In addition to the evaluative, reflective, and emotional aspects of termination, the more concrete tasks of closure must also be accomplished. Among these more concrete tasks is that of transferring any unfinished projects and/or ongoing clients to other workers.

Transferring Your Work

Much of the work that you were assigned during your fieldwork will probably be completed by the time you leave the organization. It is not uncommon, however, for students to have to leave some of their work unfinished. Work with clients is especially likely to extend beyond your field placement. A most important aspect of terminating from your fieldwork is ensuring that a smooth transition is made in any unfinished work as you leave the organization. Tasks and projects that are incomplete as well as clients who need ongoing assistance must be transferred to staff members who can bring the work to completion. You and your supervisor will plan well in advance which clients and projects are likely to need ongoing attention after you leave, thus requiring transfers to new workers. For those clients who will be continuing with new workers, it would be helpful for you to write transfer summaries to ensure that the new worker has all the information necessary to provide continuity of care. Transfer summaries include content similar to that in closing summaries (see Chapter 9) but serve the purpose of summarizing your work with the client so that the case can be smoothly transferred to a different worker for continued services. Also, it is best that you introduce clients to their new workers if at all possible before you leave. These introductions might be followed by a short meeting of the three of you together. This meeting should focus on a brief summary of your work together and a general overview of the issues that still need to be addressed. Most important, the meeting is an opportunity for the client and the new worker to begin building a relationship. Your role in this meeting is to support and encourage this new relationship, providing a smooth transition for the client from one worker to another.

As this chapter has emphasized, leaving your field site can be emotionally difficult. Many students find it especially difficult when they have to leave important components of their work midstream. Unresolved issues with clients and unfinished administrative projects can add to a student's reluctance to end the placement and create the feeling of lack of closure. In such situations, students may be especially tempted to maintain their relationship with the agency in order to see certain cases and/or projects through to completion. If your thoughts are leaning in this direction, discuss them with your field supervisor and faculty liaison. Consider carefully your motives for staying involved. Although in some cases it may be appropriate to extend the placement for a short period of time in order to finish a certain specific task, in most cases it is best for students to make a clear termination at the appointed time. The nature of human services work is such that it is often difficult to bring every task to complete closure by any particular predesignated date. Therefore, it is sometimes necessary to let go of your work, passing it to other workers to be brought to a close.

◆ EXERCISE 12.10 PERSONAL REFLECTION: OBSERVATION OF SELF AND OTHERS

What has your emotional reaction been to letting go of your responsibilities in the agency and transferring your responsibilities to others? What particular concerns or difficulties are you having in letting go of certain clients or responsibilities?

Are you tempted to try to stay in touch with any particular clients or to continue with any particular projects? If so, why?

What are your thoughts about the appropriateness of any continued contact following the end of your internship? Discuss your thoughts about this with your supervisor.

Termination Rituals

Rituals are a common and effective part of saying good-bye in many cultures and in organizations and groups, including many human service agencies. Often, rituals seem to provide comfort because they offer structure during emotionally intense situations (Fortune, 1995). Whether you have worked with groups of clients or individuals, it is likely that there will be certain ritualistic components of your leave-taking. A client might give you a small gift, a child might make you a going-away card, a group that you have co-led might mark your leaving with a going-away party, or your coworkers might take you to lunch. Such events are important ways of communicating feelings and bringing a more ceremonial or celebratory close to the experience. Rituals can be spontaneous and brief, as when a client quickly gives you a word of thanks and a hug before walking out the door, or they can be lengthy and planned, as when a group plans a potluck lunch as a going-away surprise. You, too, might want to introduce some simple elements of ritual into your ending. Writing thank you notes or finding other ways of expressing appreciation, for example, might be especially appropriate rituals as you leave your co-workers and supervisor. By the same token, brief good-bye notes to clients in which you recognize their special strengths and express your pleasure in having worked with them might also be appropriate. In any case, you might find that rituals, initiated by yourself and/or others, bring greater depth and meaning to your termination experience as well as a greater sense of closure as you say good-bye.

A Student's Reflections on Rituals

My last day of group my supervisor did a really neat thing. We focused a lot on the fact that this was my last time with the group. He used the Gestalt idea of having everyone say good-bye to me by expressing their appreciations and regrets. I also said good-bye to all of them and to him by expressing appreciations and regrets. Everyone got a piece of drawing paper and crayons. They were asked to draw a going-away gift they would like to give me. One boy gave me a new car because he has seen what a rusted-out heap mine is. Another boy drew me a picture of himself. My supervisor planned this because a lot of the guys have trouble talking about their feelings. Most of them have had a lot of people come and go in their lives without any way of talking about it. He and I also got to model for them that guys can talk about their feelings, and it's OK. I'll have to say that even though this was planned mainly because it was good for them, it was also good for me. It was one of the most memorable experiences of my entire internship.

 ◆ **EXERCISE 12.11 SYNTHESES: LINKING KNOWLEDGE AND EXPERIENCE**

What is your reaction to the idea of using rituals in terminating with clients and/or co-workers in your internship?

What rituals, if any, are being planned to mark your leaving the agency?

What kinds of rituals would you find meaningful and appropriate? Are there any rituals that you would like to initiate as part of your saying good-bye to clients and/or co-workers?

Conclusion

Leaving your internship involves many of the same processes as terminating a professional helping relationship. Evaluating your performance, reflecting on the experience, working through the emotions involved in saying good-bye, and planning your leave-taking are important parts of this process. As you work on terminating productively from your internship, you will be working simultaneously on a parallel process of helping your clients terminate their relationships with you. The content and exercises of this chapter have been directed toward helping you to better understand the termination process and to gain insight into how to apply its guiding principles to the specific tasks involved in leaving your internship.

FOR YOUR E-PORTFOLIO

As your internship ends, what do you most appreciate about your internship experience? What regrets do you have about your internship? What are your feelings, wants, and needs as you leave your internship? What do you feel you need to do for your continued personal and professional growth?

References

Alle-Corliss, L., & Alle-Corliss, R. (1998). *Human service agencies: An orientation to fieldwork.* Pacific Grove, CA: Brooks/Cole.

Baird, B. (1999). *The internship, practicum, and field placement handbook: A guide for the helping professions* (2nd ed.). Upper Saddle River, NJ: Prentice-Hall.

Barker, R. (1996). *The social work dictionary.* Washington, DC: NASW Press.

Bogo, M., & Vayda, E. (1995). *The practice of field instruction in social work: Theory and process.* Toronto: University of Toronto Press.

Brill, N. (1995). *Working with people: The helping process* (5th ed.). White Plains, NY: Longman.

Brown, J., & Pate, R. (1983). *Being a counselor: Directions and challenges.* Belmont, CA: Brooks/Cole.

Chiaferi, R., & Griffin, M. (1997). *Developing fieldwork skills: A guide for human services counseling, and social work students.* Pacific Grove, CA: Brooks/Cole.

Cormier, L., & Hackney, H. (1993). *The professional counselor: A process guide to helping* (2nd ed.). Needham Heights, MA: Allyn & Bacon.

Cournoyer, B. (2005). *The social work skills workbook* (4th ed.). Belmont, CA: Thomson Brooks/Cole.

Faiver, C., Eisengart, S., & Colonna, R. (1995). *The counselor intern's handbook.* Pacific Grove, CA: Brooks/Cole.

Fortune, A. (1995). Termination in direct practice. In R. Edwards & J. Hopps (Eds.), *Encyclopedia of social work* (pp. 2398–2404). Washington, DC: NASW Press.

Kadushin, A. (1976). *Supervision in social work.* New York: Columbia University Press.

Kaiser, T. (1997). *Supervisory relationships: Exploring the human element.* Pacific Grove, CA: Brooks/Cole.

Matorin, S., Monaco, G., & Kerson, T. (1994). Field instruction in a psychiatric setting. In T. Kerson (Ed.), *Field instruction in social work settings* (pp. 159–180). New York: Haworth Press.

Okun, B. (2002). *Effective helping: Interviewing and counseling techniques* (6th ed.). Pacific Grove, CA: Brooks/Cole.

Patterson, L., & Welfel, E. (2000). *The counseling process* (5th ed.). Stamford, CT: Thomson.

Royse, D., Dhooper, S., & Rompf, E. (1999). *Field instruction: A guide for social workers* (3rd ed.). New York: Addison Wesley Longman.

Simon, E. (1999). Field practicum: Standards, criteria, supervision, and evaluation. In H. Harris & D. Maloney (Eds.), *Human services: Contemporary issues and trends* (2nd ed., pp. 79–96). Boston: Allyn & Bacon.

Sweitzer, H., & King, M. (2004). *The successful internship: Transformation and empowerment.* Belmont, CA: Brooks/Cole-Thomson.

Thomlison, B., Rogers, G., Collins, D., & Grinnell, R. (1996). *The social work practicum: An access guide.* Itasca, IL: Peacock.

Wilson, S. (1981). *Field instruction: Techniques for supervisors.* New York: Free Press.

Young, M. (1998). *Learning the art of helping: Building blocks and techniques.* Upper Saddle River, NJ: Prentice-Hall.

Chapter 13

Planning Your Career

For most students the internship is completed near the end of their academic programs and often serves as a culminating experience in which students apply what they have learned and transition into their careers or further education. As you complete your internship, you may be trying to figure out where to go from here in your career development. There are some predictable aspects in the career development process, but no two people experience it in exactly the same way. Some students have been highly focused on a single goal for many years and ultimately enter the career they have long pursued. Others entered a human services program without a clear goal but with the conviction that they wanted to pursue a career in the helping professions. Both career paths are equally valuable, and both can lead to a successful outcome and career satisfaction.

As you complete your internship, your most pressing question might very well be, "What's next?" It is important to realize that this is not just a reflection of your relative inexperience in the field. "What's next?" is a recurring question for all growing professionals. Routinely asking yourself this question helps to ensure that you continue to learn, grow, and develop rather than stagnate professionally. To answer this question at any given time, it is useful to consider where you have been and where you are, that is, what questions you have at least tentatively answered thus far about your career and what questions remain unanswered. This chapter helps you to reflect on these and other questions as you plan the next steps in building your career.

Where Are You Now? Thinking About the "Answered Questions"

Through your human services program you have no doubt learned a great deal about yourself and the human service profession. As a result, you probably have answered many questions about yourself and your career interests and goals. Even so, it is wise to recognize that all answers to such questions are tentative, based on limited experience and an understanding that is based on your sense of yourself at a specific time in your own development. As you progress through your career, questions that were once "answered" for you can cycle back as unanswered questions once again.

Bearing these limitations in mind, it is useful to think about your answered questions at this point in your development. As you consider the following questions, think about what you have learned from your field experiences, course work, volunteer work, human service jobs, service-learning projects, and any other direct encounters you have had with the human service field. From these experiences, what have you learned about:

- yourself?
- the field of human services?
- the various populations you might work with?
- the various services you might provide?
- your strengths and weaknesses?
- what you enjoy doing? don't enjoy doing?

Pondering these and similar questions helps you figure out what you have accomplished so far in your learning process. Identifying what you have learned about human services and what you have learned about yourself are two important components of this vital step.

 ◆ **EXERCISE 13.1 PERSONAL REFLECTION: OBSERVATION OF SELF AND OTHERS**

In the space provided, make some notes responding to the bulleted questions above.

I have already taken so many directions in thinking about my career that I mostly end up confused when I reflect on them. I started as an English major but I switched to human services late in my sophomore year. I feel pretty confident that human services is for me but I still haven't really found my niche. So far I know that I want to be in a job where I can make a difference in the world but it is hard to plan a career around anything that broad. I am a good writer and have very good organizational skills. When I have done direct service work, I've felt really overwhelmed and bogged down in the very complicated situations of clients' lives. My internship has given me a glimpse of administrative work and I am thinking that may be more "my thing." I could see myself writing grants and creating opportunities for other people to provide direct services. I think that would be gratifying work. Trouble is I have only had a very limited exposure to the administrative side of things. But I want to learn more about it. I think I know what I want to do next but finding a way to do it is a different matter. That may be a challenge.

Where Are You Now? Thinking About the "Unanswered Questions"

Students often begin their work in human services programs with specific hopes about what they will learn. Many times there are specific professions within human services that students hope they will be exposed to or populations and social problems they hoped to learn about. In some cases, as students complete their programs, they find that they have learned a great deal but still find that some of their original questions remain unanswered or that their questions have changed as they have come to understand the field more deeply. Also an inevitable part of learning in any field is the reality that the more you learn, the more questions you have. The more you know, the more clearly you recognize the gaps and limitations of your knowledge. Taking stock of your unanswered questions is an important step in becoming a self-directed learner and growing professional. At this point, think about your unanswered questions by reflecting on questions such as:

◆ What types of experiences have I not yet encountered that I feel are important to my future growth and development?
◆ What populations do I need to learn more about?
◆ What unanswered questions do I have about myself? My interests? My abilities? My goals?
◆ What do I need to know about specific roles and positions in human services and the credentials that they require?

Reflections of this type help you determine what you need to do next in your own learning. Perhaps there has been an area of work that you have always dreamed of doing but that you have not yet been able to experience through your academic program or other activities. If that is the case, it may be time to engage in some research about that role. Or perhaps you feel that you have had a good exposure to the human services field but you have unanswered questions about yourself. Perhaps you are beginning to think of other careers rather than pursuing human service work. Maybe you have even become convinced that the human service field is not for you. Such thoughts can be very disturbing to students who have invested a good bit of time, money, and energy in a direction that does not seem right for them in the end. Nevertheless, if that's where you are, you will benefit from exploring this line of thought honestly. Many people take their human service knowledge and skills into other fields quite successfully. Still others wade through such inevitable moments of self-doubt to find rewarding careers in human services. The challenge at this point is simply to be honest with yourself about where you are.

 ◆ **EXERCISE 13.2 PERSONAL REFLECTION: OBSERVATION OF SELF AND OTHERS**

In the space provided, make some notes responding to the questions above.

Where Are You Going? Clarifying Your Career Goals

The successful completion of your internship is in itself a significant accomplishment. It is only natural as you accomplish such a goal to begin scanning the horizon for the next one. In thinking about your next step, the title of a well-known book about career development seems relevant, *If You Don't Know Where You're Going, You'll Probably End Up Somewhere Else* (Campbell, 1974). The intended message is that without a clear destination in mind, you are unlikely to end up in a place where you want to be.

As you think about your goals, perhaps the most important principle is allowing yourself the freedom to think flexibly and honestly about what you want to be doing five to ten years from now. It is easy to fall into the trap of meeting expectations that others have set for you or of following a script that you wrote for yourself at an earlier time but that no longer holds the same value or interest. As you complete your internship, it is time to take stock of your goals, not just in terms of what to do next but also in terms of where *you* want to be in the future. Each student's particular questions at this point will vary. Similarly, each student's way of answering those questions may vary as well. It may be useful to consider some examples of particular students' experiences at this stage of the career journey. Therefore throughout this chapter, we examine various students and how the principles of career planning apply to their situations.

CASE EXAMPLE 1: DEFINING SKILLS

Philip had completed all of his field opportunities in his human services program in health care settings. He had worked in an open door clinic for indigent patients, in a public health department doing community education, and in a hospital on the patient relations team. As he completed his internship, he felt most comfortable working in health care but there were no job openings in his community in that field. Due to his family situation, he did not have the flexibility to move to other locations where positions might be available. He also needed to begin working very shortly after graduation in order to meet his expenses as well as to begin paying off his college loans. Philip felt panicky as he realized that all of his experiences seemed to have led him to a dead-end—at least until a job became available somewhere— and he didn't have time to wait.

A staff member in the career center at his college encouraged him to think about his experiences in a different way. She helped Philip focus not on the settings of his previous work but on the skills he had developed there. Philip realized that his goal of working in health care had been based on his assumptions about where he would be most employable. As he thought about his skills more broadly, he became excited about the range of settings in which his skill might fit. He focused on identifying the specific skills he could take into a workplace and wrote a resume that conveyed those skills effectively. Rather than searching for a specific job setting, he shifted to focusing on using his skills in management, grant writing, administration, public relations, and presentation. This became his goal.

CASE EXAMPLE 2: FINDING A NICHE

As Evelyn completed her internship, she felt that she had learned primarily what she didn't want to do. Through her course work and field program, she had explored many different roles within the human services profession. She had worked in a school, a retirement center, and the local housing authority. She felt that she had learned a great deal but still had not found her niche within the field. Through her most recent internship at the local housing authority, she was introduced to the concept of community policing. As she talked with the officers who worked in community policing, she was very interested in their work. Since her program was completed, she was concerned that she would not have another field experience to test out her interest. She knew that she had too little exposure to the field of community policing to draw any definite conclusions about her suitability for the field. Evelyn approached the officers whom she had met through her internship to learn more about their work and to gather their ideas about how she might explore community policing more extensively. From these conversations, she secured an opportunity to volunteer with the program and to shadow the officers in their work for a brief period of time.

In these examples, the two students confronted different challenges as they thought about their goals, and each had reached a different degree of goal clarity.

Strategies that they employed to clarify their goals included consulting the career center, conducting a skills analysis, networking, and continuing career exploration.

For many students struggling to clarify their career goals, their best resource is the career center in their colleges and universities. The career center is frequently underutilized as students struggle to figure things out for themselves. If you are still feeling somewhat unclear about your career goals, a career planning professional may be able to help you by administering formal career assessment inventories and/or by providing counseling that will help you think through your interests, abilities, and options. Also, talking with a trusted faculty member, your field supervisor, or other mentor in the field can help you clarify your goals.

Evelyn's situation illustrates the value of networking. Through the work you have done in the community, you have met a number of experienced, well-established professionals whose work might hold interest for you now that you are embarking on your own career. Think about who is in your career network, and contact individuals whose work is particularly interesting to you. Most professionals find gratification in helping people enter their field and are flattered that others are interested in their work. Connecting with people in your network, learning more about what they do, and talking with them about your interests not only helps you to clarify your goals but also may provide leads to specific job openings.

From your experience in your human service program, you know how indispensable direct experience is in learning about the field and clarifying career directions. Continuing to volunteer in your community can extend your learning beyond the time span of your academic program. Through volunteering, you can continue to learn about human services roles, services, and populations while expanding your skills. Through this process, your career goals can become clearer and your employability can grow as your skills and knowledge increase. In some situations, volunteering can even lead to paid employment in the volunteer organization.

A Student's Reflections on Career Planning

I have been working at a group home for the past two years while I've been in school. I can continue working there, I'm sure. When I think about how I go about my job there now as compared to two years ago, I realize how much I've learned in this major and on the job. I have so many more skills than I did and I also have a much deeper understanding of the program's mission and goals. The safe thing and the easiest thing is to just keep working where I am, doing what I do, but I don't think that would really be best. I want to talk with my supervisor about opportunities for me to progress maybe into having more responsibility and more pay. Since this is a group home that has lots of different homes throughout the state, there might be opportunities in other areas to become a house manager or program co-ordinator. I think I owe it to myself after all my hard work to look at other kinds of jobs too, but the fact is I'm pretty happy where I am. I don't have to be in a hurry fortunately, but I do want to look around at what other options may be out there for me.

Achieving Your Career Goals

Some human service students complete their internships with fairly clear career goals. For these students, their questions are about the strategies they might use to accomplish their goals. Again we explore some student examples to illustrate the range of scenarios that exist even when students are clear about their goals.

CASE EXAMPLE 3: FINDING THE RIGHT FOCUS

Since she was in high school, Charlene had an interest in music therapy. She saw this career potentially as the perfect method of blending her interest in helping others with her talent as a pianist. In college she pursued course work in both music and human services. When she selected her internship, she indicated to her faculty liaison that she would like a placement in music therapy. Her professor explained to her that although music therapy might be a part of some human service programs, it was unlikely that she could secure an internship that focused exclusively on this role.

In researching possibilities, Charlene found an in-patient behavioral medicine unit in the local hospital that was interested in having an intern. Charlene believed strongly in the power of music to help people deal with their emotions and was excited to have a venue where she might try out her ideas. As she began her work on the unit, she expressed her desire to participate in music therapy. Her supervisor explained that music therapy was not a standard part of their treatment program and that there was no one on staff with expertise in this area. Her supervisor further explained that this was a pretty uncommon position in mental health care because the services of a music therapist would not be reimbursable by insurance companies or other third party payers. Therefore in most instances, the treatment centers could not afford to have them on staff. Not to be discouraged, Charlene volunteered to plan activities for the patients that would draw upon music to help them explore their emotions.

She received very positive feedback from staff and patients about her work in this area, but through becoming involved in the overall operations of the unit, she had also learned about individual therapy, group work, and family counseling. Although these areas were interesting to her, none offered the creative expression that she sought in her music therapy work, so she remained committed to this goal. By the end of her internship, she knew that she needed to do some in-depth research about music therapy and whether this was a realistic goal for her. Through her research, she located some music therapists in a nearby urban area who were willing to talk with her about their work. She learned that they worked fairly independently, contracting individually with various treatment programs for their services. Although the autonomy of this arrangement sounded appealing, she was not at all sure that this would offer her the income security that she needed. She began to think about how she might focus her career more broadly and integrate music therapy into this broader role.

CASE EXAMPLE 4: FINDING A SPECIALIZATION

Carlos grew up in a working class community where the dominant social value was helping one another. In his family and neighborhood, group support and mutual assistance were the prevailing social norms. His own sense of satisfaction and identity rested on seeing himself as a helper. He entered the human service program knowing this much about himself but not having a clear career goal. While in college, he held jobs and volunteered in several different human service settings. He also completed service-learning projects, practica, and internships that were required parts of his human service program. He had worked with children, older adults, and people with disabilities. He had his most gratifying experiences working with older adults but enjoyed all of the various roles to varying degrees.

As he completes his internship, Carlos wants to get a graduate degree in order to progress to higher level roles in his field as well as to increase his income. He had hoped to go straight to graduate school upon completing his undergraduate degree, but he is confused about what type of graduate degree to pursue. How does he know what degree to pursue if he doesn't know exactly what he wants to do? He is afraid that if he doesn't go to graduate school right away he won't go at all, and he knows that in five years he wants to be earning a reasonably good living working in human services. He feels pressure to initiate graduate school applications but he is unsure what specific field to pursue. Also, he worries that he may not be able to pay for graduate school, and he certainly doesn't want to invest his time and money in a more specialized degree that he might not be happy with.

CASE EXAMPLE 5: FINDING A WAY

Although Dori began her associate degree program in human services uncertain of what she wanted to do, she had decided by the end of her internship that she wanted to pursue a career as a school counselor. In one of her earlier field placements in her program, she had worked in a nonprofit agency that provided counseling services for children and their families. She had been very interested in this work and followed up a year later by completing an internship with a school counselor. From this experience, she felt that being a counselor in an elementary school was the perfect fit for her. She enjoyed the school environment, valued the classroom guidance opportunities that the setting afforded, and felt that she could make a positive difference by working with relatively young children and their parents in this context. She knew that she needed a graduate degree in school counseling to do this kind of work. At times, she felt overwhelmingly discouraged by the fact that she had two more years of college to complete as well as a graduate degree. Although her goal was clear, the steps in achieving that goal were not at all clear. As her two-year program comes to an end, the question "what next?" fills her with worry that she might never achieve her goals.

CASE EXAMPLE 6: FINDING A JOB

Throughout his human service program, Anwar had an interest in working with juvenile offenders. During his internship, he had the opportunity to work with Cortland County Friends of Youth where he assisted with the juvenile restitution program. By the time he had completed his internship, he had a clear goal to find similar employment. His question was how to find a job. There were no available positions in his internship site so where should he turn next?

 ◆ **EXERCISE 13.3 PERSONAL REFLECTION: OBSERVATION OF SELF AND OTHERS**

In what ways are the student examples above similar to your own experience as you conclude your internship? Think about each of these students individually. What suggestions do you have for how they might achieve their goals? Among your suggestions are their ideas that might be useful to you in your own situation right now?

Basic Steps in Career Development

There are, of course, many paths open to each of the students above as they work toward their goals. While there is not one right approach for anyone, there are three basic steps to follow in achieving your career goals. They are:

Step 1: Clarifying long-term goals as distinct from short-term goals or intermediate goals.

Step 2: Doing your research (for a job and/or an advanced academic program).

Step 3: Conducting a job search: Developing a resume, writing effective letters, and developing interviewing skills; maintaining a positive outlook.

Each of these steps, and how each applies to the various students introduced above, is discussed in detail in the remainder of this chapter.

Step 1: Clarifying Long-Term Goals as Distinct from Short-Term Goals or Intermediate Goals

As you think about your future, it is most useful to think in terms of short-term goals and long-term goals. In some cases, even intermediate-term goals should be included in your plan. There are many positions in human services that are not available immediately upon completion of a two-year or four-year degree. If your ultimate goal is to secure such a position, it is necessary to identify the steps for achieving that goal. It may be necessary to acquire specific types of work experience and/or an advanced degree in order to achieve your long-term goal. The key is to develop a plan that is realistic and to stay the course in implementing the plan.

Both Dori and Carlos are students who recognize that the careers they ultimately want will not be immediately available to them. Therefore they need to set some short-term goals and intermediate-term goals that will meet their pressing needs for employment while also moving them closer to their long-term career goal. Pursuing her interest in working in a school, Dori explored employment in the local school system where she was hired as an elementary school classroom teaching assistant. While employed at the school, she continued to work toward her four-year degree through evening, on-line, and summer classes. She also became well acquainted with the two school counselors who were employed at her school and talked with them at times about her career interest. A few years later, a new position was developed in the school for a school counseling assistant. Dori enthusiastically applied for this position and was hired.

She continued working in the counseling office at the school until her undergraduate degree was completed. She enjoyed her work and had many opportunities to meet counselors from other schools as she worked in this role. When she had the opportunity, she talked with them about graduate schools they had attended and the career paths they had followed. She also researched school counseling masters program independently. She learned about some graduate programs that were built on a work-study model that would enable her to continue working while she took graduate level courses. By the time she had completed her bachelor's degree, she had already applied to a university that offered the work-study option. Her extensive experience working in education made her a strong candidate for graduate study and she was accepted. Dori was ecstatic that she could now see her long-range goal in sight as she embarked on earning her masters degree.

Carlos began to clarify his goals by researching graduate programs. He looked at programs in public administration, public health, social work, and counseling. Based on his research, he felt that the master of social work degree would probably be best for him because of its breadth. It would allow him to work in a broad range of service areas and with a broad range of populations. He especially liked that social work programs included education for both direct services and administration. While he thought he would want to concentrate on administration, he would not have to declare a concentration until his second year. In the meantime, he could continue exploring his exact interests. Although he was eager to begin his graduate work, Carlos decided that he would be wise to work a year or two to be more certain of his tentative choice to pursue social work. He would also have time to save a little money and explore financial assistance for graduate study as well. Therefore Carlos decided that his short-term goal should be finding a job that would help him learn more about his interests and allow him to test his hypothesis that social work would be a good field for him. His tentative intermediate-term goal was to earn a master's degree in social work. His long-range goal was to become an agency administrator.

A Student's Reflections on Career Planning

I come from a long line of people who have either worked in business or in technical areas like medical technology or engineering. My family has almost no understanding of what I want to do, so I have to look for support and guidance in other places. The faculty here has been wonderful for that. Now in my internship I have

continued

---

met another whole group of people who have taken a lot of interest in me and my career. They all have advice for me about what I should and shouldn't do. Since I am in the school system, I have a chance to work with school counselors, school social workers, school nurses, etc. I get to see a whole spectrum of professions in the course of the day. Before I leave, I definitely want to set up a formal interview with some of them to learn more details about their jobs and their degrees. I am open to getting a graduate degree but I don't know what in yet. I realize I am in the ideal spot to do some research, so I'm definitely going to take advantage of it. Everybody is so busy it's hard to work these conversations into the normal day so I know I will need to schedule some meetings if this is going to happen.

Step 2: Doing Your Research (For a Job or an Advanced Academic Program)

The process of developing a career plan inevitably involves research. Whether your next step is pursuing further study or searching for a job, research will be an important part of successfully achieving your goal. Both Dori and Carlos in the examples above engaged in research about academic programs. The Internet provides a convenient way to do this research. Although you will want to talk with program graduates and visit programs before making your decision about specific academic programs to pursue, program websites provide a great deal of information about the program philosophy, curriculum, faculty, and accreditation status.

It is important to note that even though a program may have interesting courses, a sound curriculum, and a highly trained faculty, it may not be the right program for the goals you have in mind. Furthermore, even programs that seem to fit perfectly with your goals and interests may not hold the proper accreditations and professional affiliations necessary to advance your career. Any position that requires licensure or certification for employment requires candidates to be graduates of accredited programs in the field. For example, positions for licensed or certified social workers require that candidates hold a degree from a program accredited by the Council on Social Work Education (CSWE). Similarly, positions for certified or licensed counselors require the candidate to hold a degree from a program accredited by the Council for Accreditation of Counseling and Related Educational Programs (CACREP).

In addition to earning the accredited degree, licensure or certification generally requires a number of supervised hours of professional practice in the field as well as passing a formal exam. Specific details of licensure and certification vary with each state. Therefore you also need to research the exact requirements for employment, licensure, and/or certification in the state(s) in which you will be pursuing employment. This information may be acquired through the websites of the certification boards in each state.

Although many students pursue graduate studies immediately after completion of a four-year degree, most do not. Many jobs in human services do not require an advanced degree, certification, or licensure. Most graduates seek employment immediately after graduation and engage in a job search in order to accomplish this goal. Research is an important component of seeking a job, just as it is with selecting a graduate program. Research is so important to the job search process at this stage that Richard Nelson Boles (2001), author of the best-selling job-hunting book, *What Color Is Your Parachute?*, suggests that you should think of yourself as a "job-researcher" rather than as a "job-seeker."

To understand the various forms of research involved in a job search, we will examine a continuation of Anwar's story. Like Dori and Carlos, Anwar decided to pursue a job immediately upon the completion of his degree. As you recall, he specifically wanted to secure a position working with juvenile offenders. Through examining Anwar's job search, you will see that research was fundamental to the entire process. As you read about his job search, note all of the forms of research it required.

CASE EXAMPLE 6 CONTINUED: RESEARCH

As Anwar began searching for a job, he turned his attention first to researching agencies where he might seek employment. He followed several paths in this research. He began by talking with agency staff members he had worked with during his internship. From them he learned about similar agencies in surrounding counties and was able to make a list of contact names, agency names, addresses, and phone numbers. He also did an Internet search that turned up a number of additional possibilities. Through talking with friends and neighbors, he learned about people who were currently employed in the field. In some cases, his contacts were able to facilitate opportunities for him to meet with these people. These meetings provided him the opportunity to learn about their organizations and to get the word out about his own interests.

He talked with the staff in career services at his college and checked the career services job postings regularly. Through checking the local telephone directory, he was reminded of some broad-based agencies, such as the local mental health program and the public health department, that might also provide services for adolescents. Anwar checked with the local Employment Security Commission where all state job openings were posted. He also researched the availability of any federal jobs in his area. In addition, he made contact with the local United Way organizations in a five-county area where he was interested in working. Through the United Way offices he was able to secure a comprehensive directory of all of the human services agencies in these counties. Through all of this research, Anwar was able to construct a personal database of all of the organizations that he would like to contact regarding employment.

Research was an important and multipronged process in Anwar's job search. There was no "one-stop shop" where he could find all of the information he needed. He drew upon informal sources of information, such as his professional network and friends, as well as formal sources, such as career services and employment websites. A number of websites exist to facilitate job searches in the nonprofit sector, including The Nonprofit Career Network, Opportunity Knocks, the Social Service Job Site, and the Human Services Career Network.

In conducting research on the organizations where you might secure employment, your research should be both broad and deep. As you begin to make direct contact with organizations, it is not enough to have only contact information. You want to know all that you can about the organization's mission and the services it provides. Most of this information and more is provided on agency websites. Agencies that do not have websites generally have print materials available, such as brochures or annual reports. You might visit the office to pick these up or ask that they be mailed to you.

Career services at your college or university will also be able to provide information about some of the organizations that you are interested in. Many career centers have relationships with alumni who have worked in various careers and settings who are eager to mentor students, sharing their knowledge, experience, and expertise. Research on successful job searches reveals that it is through such direct person-to-person contacts that most job seekers find success in the job market (Bolles, 2001).

◆ EXERCISE 13. 4 ANALYSIS

What research have you done to prepare for your job search? What research do you need to do? What questions do you want to answer? What sources of information will you draw upon in your research?

Once you have gathered sufficient information to guide your search, you will be ready to take more active steps to secure a job, transitioning from job-researcher to job-seeker.

Step 3: Conducting a Job Search

Looking for a job is hard work, requiring a number of skills and tasks that are not necessarily taught in college classrooms. These skills and tasks include developing a resume, developing interviewing skills, writing well-crafted letters, and maintaining persistence and a positive outlook over time. We examine each of these briefly.

In addition to the discussion here, the career services staff of your institution is an excellent resource for gaining the skills required in job seeking. Career services routinely offers assistance with constructing resumes and letters to potential employers. It also provides opportunities to interview and/or network with potential employers. Some career services centers even provide opportunities for videotaping yourself in a mock job interview. You can then review the videotape to critique and improve your performance. Career services often provides for credit courses or noncredit workshops on how to secure a job. In addition, it often maintains a network of alumni contacts who have indicated an interest in mentoring students or graduates as they transition into their careers.

Such services are free and easily at your disposal. For all of these reasons and more, students should avail themselves of the expertise and services of the career planning professionals at their college or university. There are also many helpful websites that focus on career planning, resume writing, job interviewing, and other career building skills. For such Internet information you might visit, for example, the website for the Human Services Career Network, Idealist.org: Action Without Borders, and The Riley Guide.

CASE EXAMPLE 6 CONTINUED: THE JOB SEARCH

With the assistance of his academic advisor and career services at his college, Anwar developed a resume. He sought feedback on his resume from several people, including his internship field supervisor and his internship faculty supervisor. After revising his resume several times, he felt confident that it presented an excellent picture of his skills and experiences. Through frequently checking with career services at his college and routinely reviewing newspaper classified ads, he was conscientious in pursuing active openings in the field.

Rather than send his resume out as a mass mailing to all of the organizations he was interested in, Anwar tried to make a personal contact at the organization, first either through a telephone call or a scheduled appointment. If there were no open positions, Anwar asked if the program director would be willing to meet with him for a few minutes so he could learn more about the organization. He developed a specific cover letter tailored to each organization to accompany his resume when he sent it out. He kept a log of all of his mailings and followed up a few days later with a phone call to ensure it was received. He followed up all formal job interviews or exploratory conversations with letters as well.

Anwar realized that a successful job search generally takes months so he began his search well before graduation and worked on maintaining realistic expectations. He did not expect overnight success and was usually able to maintain his energy and optimism as the search progressed.

Developing a Resume

Your resume is a one- to two-page summary of your experience, skills, and educational background. It is one of your major marketing tools in selling yourself to a potential employer. Although the resume is vitally important to a successful job search, its importance can be exaggerated. An excellent resume will not get you a job. Employers hire people, not resumes. Most employers rely heavily on the personal interview as they determine who they will hire. The resume and accompanying cover letter (discussed later) function to help the employer decide who to interview. Your goal is to write a resume that effectively conveys your skills and experience, persuading the potential employer that you should be interviewed for the job.

Although an excellent resume cannot get you a job, a weak resume can certainly lose a job for you. Because most employers review many resumes for a given position, they must review them critically. When a pool of sixty candidates for a position must be winnowed down to three who will be interviewed, employers must look for reasons to reject many of the resumes as well as reasons to retain a few for further consideration. Misspellings, disorganization, lack of clarity, or confusing, crowded formats can give employers easy reasons to eliminate you from the pool. Therefore focusing on the basics of good writing, clear presentation of ideas, and an easily grasped structure of the document is essential to pass even the most cursory screening of the resumes. Having passed this hurdle, your resume can then be reviewed for significant content, that is, your experience, skills, and educational background.

Most experts on resume writing agree that an effective resume summarizes your objective, your experience, your skills, and your accomplishments. The top of page one will carry basic contact information—your name, address, phone number, and e-mail address. Directly under this identifying information is your objective. Your objective states clearly what you are hoping to achieve through your job search. Although there is some disagreement about the importance of including the objective, most resources on resume writing stress its important as a key piece of information that helps the employer interpret your resume. The objective is the organizing principle for the rest of the document, bringing your skills and experience into the proper focus (Krannich & Krannich, 2000). For example, your objective might be, "To provide direct care for people with disabilities through using my demonstrated skills in life skills education, behavior management, and treatment plan implementation." The reader will consider subsequent information you provide about your skills and previous positions in the context of this objective.

There are many possible formats for the remainder of the document. The two most common are the reverse chronological and the functional approaches. Most prevalent is the reverse chronological resume. Using this format, you list your most recent professional experience first, followed by your previous experiences in descending chronological order, finally listing your first experience last. Skills, responsibilities, and achievements are then listed under each job heading as they pertained to your work in that organization. Functional resumes use skills rather than previous employers as the primary organizing principle.

In a functional resume, you identify three to five broad skill headings and under each you list a number of specific skills. For example, drawn from the job objective above, the skill headings for a functional resume might include life skills education, behavior management, and treatment plan implementation. The heading for the skills section might be labeled "Skills" or "Qualifications." In a functional resume, work experiences are listed briefly following the skills section of the resume. The order of the work experiences can either be reverse chronological, or they can be organized functionally. A functionally organized work experience section lists first the position you have held that is most closely related to your objective and lists last the position that is least closely related to that objective (Dowd & Taguchi, 2004).

Regardless of format, it is important that your resume include a clear description of your skills, using action words to lead each statement. Examples of such skill statements include: "Provided direct care to five adolescents with autism in a group home setting" or "Designed and implemented an arts activity program for older adults with limited mobility." Identifying your most important skills and conveying them in your resume is perhaps the most important task in creating an effective resume. In fact, most of the content of your resume is the identification of skills.

Closely related to skills are your accomplishments. Including a "Special Accomplishments" section in your resume gives you the opportunity to highlight particularly impressive achievements. This section, should you choose to include it, focuses on particularly impressive efforts on your part. These entries might focus on unusual initiatives or particularly successful results of your efforts. Using numbers to quantify specifics of your accomplishments helps to paint a clearer picture of your work. Examples of accomplishment statements might include:

◆ Initiated and organized a campus-wide fundraiser that yielded over $12,000 for sexual assault services.

◆ Initiated, developed, and led an innovative summer camp program for children of diverse backgrounds to learn about one another's cultures. The camp enrolled 200 children and was granted the Award of Excellence by the local Human Relations Council.

◆ Chaired a conference planning task force that organized a three-day national conference with over 300 participants and 25 presenters. Oversaw a budget exceeding $20,000 dollars as well as a student staff of 30 people and 6 committees. Succeeded in yielding a profit of $3,000 as well as excellent evaluations of conference quality.

Through briefly listing three to five personal accomplishments, you can succinctly represent to potential employers special abilities and traits that you can bring to their organization (Dowd & Taguchi, 2004; Krannich & Krannich, 2000).

After outlining your objective, work history, skills, and accomplishments, you might consider your resume complete. In some cases, however, you might choose to include more information. A frequent question has to do with whether resume writers should include personal information or information regarding hobbies. In general, experts recommend omitting this information in most cases. Exceptions are made if you have interests or personal information that is particularly relevant to the position. If, for example, you are applying for a position in an adventure-based counseling program and your hobbies include rock climbing, sailing, and camping, that is relevant information regarding your potential comfort with the position (Krannich & Krannich, 2000).

Another common question is whether to include references on the resume. The general recommendation is that at most you might include a statement such as "References provided upon request." The best practice is to select references who can provide the most relevant information related to a particular job for which you are applying. At the time of your interview, it is wise to take a list of these references so that you can provide them should they be requested (Krannich & Krannich, 2000).

Once your resume is complete, you need to make decisions about font, format, and paper. It is generally recommended that you use a quality paper of 20 lb. to 50 lb. bond in a color that is neutral and conservative, such as white, gray, off-white, or beige. Heavy, textured papers are generally not recommended. The font you select should be 12 point, traditional, conservative, and easily legible. The ink color should be black. The format of your resume should allow sufficient white space so that the page does not look crowded and the reader is not overwhelmed with information. It should not be your goal to put as many words as possible on each page (Krannich & Krannich, 2000).

E-mailing or faxing resumes should not be done unless explicitly requested by the employer. There are many Internet sites that are set up for job posting and resume posting. You may have questions as to whether this is a good idea for you. Certainly if the service is free, you may have little to lose. You should also be aware, however, that using the Internet to find a job is a notoriously poor method of job searching. Bolles (2001) estimates that only 1 percent of job seekers secure a job using this method. Another poor method is mailing out resumes at random. Only 7 percent or fewer find jobs through broad, untargeted mailings of resumes (Bolles, 2001). Remember that your resume is a tool and knowing the ways to use this tool enhances its effectiveness.

Writing Effective Letters

Within your job search, there are many occasions to write letters. Each resume you send to a potential employer should be accompanied by a cover letter. The cover letter is so called because it is delivered on top of the resume. It should not be stapled to the resume. Each cover letter should be targeted to the particular position that you are interested in. You should not mass produce a standard one-size-fits-all cover letter. Your letter should be addressed to a particular individual if at all possible rather than "To Whom It May Concern." Phone the agency if necessary to secure the name of the individual to whom the letter should be addressed. In almost all cases, the cover letter should be no more than one page and should follow the format of a standard business letter. Centered at the top of the page in a letterhead format, you should include all methods of communicating with you including name, address, phone number, and e-mail address. You want to make it as easy as possible for the employer to contact you (Stuenkel, 2001).

The primary purpose of the cover letter is to highlight the connections between your experiences and skills and the needs of the employer. This letter functions to prepare the reader to see your resume in light of the position. In content, your cover letter should state facts and accomplishments and relate them to the position that you are interested in. A cover letter might state, for example, "In my most recent position, I provided case management services for a caseload of 35 clients and chaired a community task force that was charged with improving interagency communication. My background seems consistent with your needs for a professional to provide case management services and to lead community initiatives." Your goal is to write a letter that will persuade the reader to examine your resume and to invite you for an interview. In the closing paragraph, you should directly request an interview (Stuenkel, 2001).

In addition to the all-important cover letter, there are many occasions within a job search that you should respond to with a letter. These occasions include informational interviews, job interviews, withdrawing from a search, responding to a rejection, and accepting a job offer (Krannich & Krannich, 2000). Although these missives are sometimes referred to as thank-you notes, you should think of most as standard business letters that include, among other content, an expression of thanks. These letters should be written in a standard business format and should include additional information beyond the expression of appreciation. Following each interview, you should write a letter in which you not only express appreciation for the opportunity to interview but also emphasize specific points related to your suitability for the position. Such letters should be sincere and express your continued interest in the position as well as your confidence that you can do the job (Stuenkel, 2001).

Thank-you notes per se are more appropriate following purely informational interviews. When an individual has granted you some time for the express purpose of sharing information that might be helpful in your career, this was indeed an act of generosity and kindness. A thank-you note is therefore always in order. Job interviews, however, are business meetings in which both you and the employer are working to determine whether there is a suitable match between your qualifications and the organization's needs. Therefore a business letter is a much more appropriate response. Follow-up letters of this sort should be thought of as a standard part of your job search, equal in importance with the cover letter and resume.

 ◆ **EXERCISE 13.5 SYNTHESIS: LINKING KNOWLEDGE AND EXPERIENCE**

If you have not yet developed a resume and written some practice cover letters, begin work on this immediately. Contact your school's career services office for assistance and feedback on your work.

Developing Interviewing Skills

At last, the moment you've been waiting for! You have been asked to interview for a position. This is the opportunity for the employer to get to know you as a person as well as your qualifications. Likewise, this is your opportunity to gather more information about the organization and the position. The importance of the interview cannot be overemphasized. It is through this conversation that the employer determines not only whether you have the appropriate skills and qualifications for the position but also whether you are a good fit with the organization personally. Although you might reasonably assume that the most well-qualified candidate gets the job, other factors come into play in the interview. Social skills, sincerity, poise, comfort level, and personal style all play a part in the chemistry of the interview. The employer, in short, wants to find a good employee. A good employee is, of course, someone who is qualified to fulfill the requirements of the position. Equally important, however, is the fact that a good employee is also someone who works well with others, puts in a full day's work, takes initiative, and has an interest in the position beyond the income it will produce. Therefore your interviewer will quite understandably try to get a sense of who you are as a person and will most likely take the interview beyond discussion of your skills and qualifications.

You should arrive for the interview a few minutes early and be polite and gracious to everyone with whom you have contact. As you know from your fieldwork, professional dress in most human services organizations is generally more casual than is the case in corporate settings. You can get a sense of the typical attire of the organization from talking with someone you know who works there or from watching employees enter and leave the building. If you do not have access to this type of information, you might consider asking your interviewer, once you have settled upon a time and date for your interview, about typical dress within the organization. The rule of thumb in selecting your attire for the interview is to err on the side of being slightly more formal than the dress of the typical employee. Also it is wise to dress conservatively. Both you and your clothing must at minimum be neat and clean. Your clothing should not call attention to itself. You want the interviewer to remember you, not what you were wearing. You should greet the interviewer with a warm smile and a firm (not limp or bone-crushing) handshake (Bolles, 2001).

In the interview, you want to convey familiarity with the organization's mission and services as well as at least a basic understanding of the position in question. Therefore, as you approach your interview, you might need to do more research about the organization. In preparing for the interview, you should also work on a description of yourself that you can convey to your employer in a brief statement of just a few minutes. This statement, of course, should be delivered in a conversational manner and should focus on your background, your strengths and skills with supporting examples or details, and your goals.

Many interviewers open with a fairly open-ended question such as "Tell me about yourself," which will give you the opportunity to convey what you feel is important about yourself (Getting Your Ideal Job, 1999). You do not want to squander this open-ended opportunity to make your case with vague statements about yourself or extraneous information that is not relevant to the job. As the interview moves along, it is likely to become more structured, perhaps giving you less opportunity to deliver the message you want to deliver. As you talk about yourself throughout the interview, be sure to include specific examples of your skills rather than just talk in generalities. An interviewer might note, for example, that it is important for all employees to maintain their documentation in a timely fashion in order for agency funding to be secure. Rather than responding with a general statement that you can handle that requirement, you should follow up such an assertion with an example of how you have handled this or similar requirements successfully in other situations.

As you respond to questions about previous experiences and employers, always maintain a positive approach. Do not criticize previous employers or find fault with colleagues. Interviewers might assume that you will bring this same critical disposition into their workplace. Throughout the interview maintain a strong, clear voice that is loud enough to be heard but not overpowering. Of course, using proper grammar and maintaining good eye contact are basic ingredients of professional behavior and should also be present throughout the interview. Listening carefully to the interviewer is also important in order to understand the questions you are being asked and their context. Approaching the interview with the question, "What can I do for this employer?" and framing your remarks around this question bring the proper tone and focus to the interview. Because many candidates tend to focus on either just getting the job or what the organization can do for them, candidates who focus on meeting the organization's needs often stand out (Bolles, 2001).

Although you might think of the interview as a time when you answer questions, you should be prepared to ask questions as well. Interviewers routinely ask interviewees whether they have any questions. Asking questions conveys your interest in the organization and generates information that will help you to make a good decision as to whether this position and this organization are right for you. You might have questions about the position itself or the broader organization. You might ask questions about the organization's goals for the future. You might inquire about the organization's greatest strengths and assets as well as its greatest challenge. You might ask the interviewer what he or she has found to be most rewarding about working in the organization. In short, you should target your questions to help you get a sense of the organization and what it might be like to work there. Additionally, you want your questions to convey genuine interest without being invasive or negative in tone.

As the interview closes, it is appropriate to convey your appreciation for the opportunity to interview and to learn more about the position and the organization. It is also appropriate to ask any questions you might have about when a decision might be made and an approximate time that you might hear back from the interviewer. Closing with the same firm handshake and smile that you began with and conveying your continued interest in the position are recommended.

◆ EXERCISE 13.6 SYNTHESIS: LINKING KNOWLEDGE AND EXPERIENCE

Conduct at least one mock interview in which you are interviewing for a job in a human service agency. Your interviewer can be a fellow student, family member, faculty member, or career services staff member. At the end of the interview, critique yourself on your performance. Also ask for feedback from your interviewer. Repeat the process with different interviewers to get more varied feedback and to become more comfortable with the process.

A Student's Reflections on Career Planning

Today was a big day—my first interview for a real job! I interviewed for a job at the local hospital to work in patient relations. It is kind of a cross between human services and customer service. Eventually I would like to work in hospital administration and so this may be a good place to start toward that goal. I think the interview went pretty well but I do realize that I need to know more about medical settings. Some of the jargon and acronyms that the interviewer used were like a foreign language to me, like JCAHO, JCR, HIPPA, DRGs, etc. Before I have another interview I will definitely bone up on some of the issues related to health care administration so I won't feel so out of my element.

Maintaining a Positive Outlook

Conducting a job search is a time-consuming task requiring hard work, patience, and persistence. People who enter human services are typically not very interested in marketing and are often more comfortable focusing on others than on themselves. Therefore an intense process of marketing themselves is particularly challenging for many in the field. Perhaps the most important element to maintaining a positive outlook is going into the job search with realistic expectations. The exact length of time the typical job search takes fluctuates with shifts in the economy and other cycles, but it is safe to say that you should be prepared for your job search to take months. Going into the process with realistic expectations helps inoculate you against impatience and disappointment.

Added to this already challenging situation is the fact that rejection is an almost inevitable part of the job-search process. Although it is difficult to know how to interpret a rejection, you should know that it is possible to do everything right and still not get the job. There might have been a candidate with experience or qualifications that more closely fit the organization's needs. That does not mean that you did not perform well on the interview, have a weak resume, or should have revised your cover letter for the tenth time. Job searching is in its very nature competitive, which means that nobody "wins" all the time. It is useful to get what feedback you can in order to learn from it, but you should guard against blaming yourself if you do not get a particular job that you want. Many times a job candidate can do everything right and still not get the job. Be open to feedback, but do not assume that a rejection means that you made mistakes somewhere along the line. Rejection for a job does not necessarily mean that you did anything wrong at all. Employers often have tough choices to make between good candidates. There are many factors in any job search that are outside your control, so blaming yourself is not realistic in many situations.

Chapter 10, Managing Your Feelings and Your Stress, provides many helpful strategies for maintaining a realistically positive outlook. A particularly applicable strategy to the job search is developing positive self-talk. Positive self-talk involves first recognizing when you are giving yourself unrealistic, negative messages about a situation. Blocking such messages and replacing them with reality-based messages is an effective way of managing stressful situations. Rather than focusing on messages that "catastrophize" about being turned down for a job, you should construct more reality-based messages about the situation. For example, catastrophizing might involve saying to yourself, "I'll never get a job. I'm just a failure," whereas positive self-talk would involve giving yourself more realistic messages such as, "Only one person can get a given job. Most people who applied for the job were turned down. If several qualified, competent people apply, all but one must be rejected."

It is also realistic and helpful to see each job interview as a learning opportunity that can help you become better prepared for the next one. In some situations, it may even be appropriate to ask your interviewer for feedback about specific ways you might enhance your employability. Asking for feedback on your resume and your interview, as well as on your skills and qualifications can yield useful information for improvement. Asking for such feedback should be limited to those interviewers with whom you feel you established a good rapport and whose feedback you would respect and value. You might also ask for their suggestions about any other agencies and organizations that might be interested in hiring you. The key is to try to make each situation an opportunity to learn and progress.

Conclusion

As the internship ends, students have varying needs, interests, and goals. Questions about career development are frequently prominent. Completion of the internship and even completion of the human service program does not ensure that you now have clear career goals. Many graduates find that they are still in the process of clarifying their interests, career goals, and plans for further education. This is normal and reflects the fact that career development and professional identity continue to evolve throughout a lifetime. Children are often asked what they want to "be" when they grow up. As an adult, you may now have a more accurate understanding of professional life as an ongoing process of "becoming" rather than as a fixed state of "being." Growing professionals continually reflect on where they have been, where they are, and where they are going professionally, just as you have done in completing this chapter. I wish you the best on this exciting journey. I am gratified to have played a small part in that journey.

 ◆ **FOR YOUR E-PORTFOLIO**

Write a portfolio entry in which you describe your career goals and your plan for achieving them. You might consider the following questions: What are your long-term and short-term career goals as your internship ends? What would you like to be doing professionally five or ten years from today? What specific steps, if any, do you need to take in order to clarify your career goals? What are your next steps as you continue to pursue your personal and professional growth? How will you accomplish these steps? What is your timeline for the completion of various steps in this process? What do you feel you need to learn in order to accomplish your goals?

References

Bolles, R. N. (2001). *What color is your parachute? A practical manual for job-hunters and career changers*. Berkeley, CA: Ten Speed Press.

Campbell, D. (1974). *If you don't know where you're going, you'll probably end up someplace else*. Allen, TX: Thomas More.

Dowd, K. O., & Taguchi, S. G. (2004). *The ultimate guide to getting the career you want and what to do once you have it*. New York, NY: McGraw Hill.

Getting your ideal job: Networking, interviewing, and landing your job offer. (1999). San Francisco, CA: Wetfeet Press.

Krannich, R. L., & Krannich, C. R. (2000). *The savvy resume writer: The behavioral advantage*. Manassas Park, VA: Impact Publications.

Stuenkel, L. A. (2001). *From here to there: A self-paced program for transition in employment* (4th ed.). Lawrence, KS: Lawrence & Allen.

Appendix

Ethical Standards of Human Service Professionals

National Organization for Human Service Education

Preamble

Human services is a profession developing in response to and in anticipation of the direction of human needs and human problems in the late twentieth century. Characterized particularly by an appreciation of human beings in all their diversity, human services offers assistance to its clients within the context of their community and environment. Human service professionals, regardless of whether they are students, faculty, or practitioners, promote and encourage the unique values and characteristics of human services. In so doing, human service professionals uphold the integrity and ethics of the profession, partake in constructive criticism of the profession, promote client and community well-being, and enhance their own professional growth.

The ethical guidelines presented are a set of standards of conduct that the human service professional considers in ethical and professional decision-making. It is hoped that these guidelines will be of assistance when the human service professional is challenged by difficult ethical dilemmas. Although ethical codes are not legal documents, they may be used to assist in the adjudication of issues related to ethical human service behavior.

Human service professionals function in many ways and carry out many roles. They enter into professional-client relationships with individuals, families, groups, and communities, who are all referred to as "clients" in these standards. Among their roles are caregiver, case manager, broker, teacher/educator, behavior, changer, consultant, outreach professional, mobilizer, advocate, community planner, community change organizer, evaluator, and administrator. The following standards are written with these multifaceted roles in mind.

The Human Service Professional's Responsibility to Clients

Statement 1. Human service professionals negotiate with clients the purpose, goals, and nature of the helping relationship prior to its onset, as well as inform clients of the limitations of the proposed relationship.

Statement 2. Human service professionals respect the integrity and welfare of the client at all times. Each client is treated with respect, acceptance, and dignity.

Statement 3. Human service professionals protect the client's right to privacy and confidentiality except when such confidentiality would cause harm to the client or to others, when agency guidelines state otherwise, or under other stated conditions (e.g., local, state, or federal laws).

Statement 4. If it is suspected that danger or harm may occur to the client or to others as a result of a client's behavior, the human service professional acts in an appropriate and professional manner to protect the safety of those individuals. This may involve seeking consultation, supervision, and/or breaking the confidentiality of the relationship.

Statement 5. Human service professionals protect the integrity, safety, and security of client records. All written client information that is shared with other professionals, except in the course of professional supervision, must have the client's prior written consent.

Statement 6. Human service professionals are aware that in their relationship with clients, power and status are unequal. Therefore they recognize that dual or multiple relationships may increase the risk of harm to, or exploitation of, clients, and may impair their professional judgment. However, in some communities and situations, it may not be feasible to avoid social or other nonprofessional contact with clients. Human service professionals support the trust implicit in the helping relationship by avoiding dual relationships that may impair professional judgment, increase the risk of harm to clients, or lead to exploitation.

Statement 7. Sexual relationships with current clients are not considered to be in the best interest of the client and are prohibited. Sexual relationships with previous clients are considered dual relationships and are addressed in Statement 6.

Statement 8. The client's right to self-determination is protected by human service professionals. They recognize the client's right to receive or refuse services.

Statement 9. Human service professionals recognize and build on client strengths.

The Human Service Professional's Responsibility to the Community and Society

Statement 10. Human service professionals are aware of local, state, and federal law. They advocate for change in regulations and statutes when such legislation conflicts with ethical guidelines and/or client rights. Where laws are harmful to individuals, groups, or communities, human service professionals consider the conflict between the values of obeying the law and the values of serving people and may decide to initiate social action.

Statement 11. Human service professionals keep informed about current social issues as they affect the client and the community. They share that information with clients, groups, and community as part of their work.

Statement 12. Human service professionals understand the complex interaction between individuals, their families, the communities in which they live, and society.

Statement 13. Human service professionals act as advocates in addressing unmet client and community needs. Human service professionals provide a mechanism for identifying unmet client needs, calling attention to these needs, and assisting in planning and mobilizing to advocate for those needs at the local community level.

Statement 14. Human service professionals represent their qualifications to the public accurately.

Statement 15. Human service professionals describe the effectiveness of programs, treatments, and/or techniques accurately.

Statement 16. Human service professionals advocate for the rights of all members of society, particularly those who are members of minorities and groups at which discriminatory practices have historically been directed.

Statement 17. Human service professionals provide services without discrimination or preference based on age, ethnicity, culture, race, disability, gender, religion, sexual orientation, or socioeconomic status.

Statement 18. Human service professionals are knowledgeable about the cultures and communities within which they practice. They are aware of multiculturalism in society and its impact on the community as well as individuals within the community. They respect individuals and groups, their cultures and beliefs.

Statement 19. Human service professionals are aware of their own cultural backgrounds, beliefs, and values, recognizing the potential for impact on their relationships with others.

Statement 20. Human service professionals are aware of sociopolitical issues that differentially affect clients from diverse backgrounds.

Statement 21. Human service professionals seek the training, experience, education and supervisions necessary to ensure their effectiveness in working with culturally diverse client populations.

The Human Service Professional's Responsibility to Colleagues

Statement 22. Human service professionals avoid duplicating another professional's helping relationship with a client. They consult with other professionals who are assisting the client in a different type of relationship when it is in the best interest of the client to do so.

Statement 23. When a human service professional has a conflict with a colleague, he or she first seeks out that colleague in an attempt to manage the problem. If necessary, the professional then seeks the assistance of supervisors, consultants or other professionals in efforts to manage the problem.

Statement 24. Human service professionals respond appropriately to unethical behavior of colleagues. Usually this means initially talking directly with the colleague and, if no resolution is forthcoming, reporting the colleague's behavior to supervisory or administrative staff and/or to the professional organizations to which the colleague belongs.

Statement 25. All consultations between human service professionals are kept confidential unless to do so would result in harm to clients or communities.

The Human Service Professional's Responsibility to the Profession

Statement 26. Human service professionals know the limit and scope of their professional knowledge and offer services only within their knowledge and skill base.

Statement 27. Human service professionals seek appropriate consultation and supervision to assist in decision-making when there are legal, ethical, or other dilemmas.

Statement 28. Human service professionals act with integrity, honesty, genuineness, and objectivity.

Statement 29. Human service professionals promote cooperation among related disciplines (e.g. psychology, counseling, social work, nursing, family and consumer sciences, medicine, education) to foster professional growth and interests within the various fields.

Statement 30. Human service professionals promote the continuing development of their profession. They encourage membership in professional associations, support research endeavors, foster educational advancement, advocate for appropriate legislative actions, and participate in other related professional activities.

Statement 31. Human service professionals continually seek out new and effective approaches to enhance their professional activities.

The Human Service Professional's Responsibility to Employers

Statement 32. Human service professionals adhere to commitments made to their employer.

Statement 33. Human service professionals participate in efforts to establish and maintain employment conditions that are conducive to high quality client services. They assist in evaluating the effectiveness of the agency through reliable and valid assessment measures.

Statement 34. When a conflict arises between fulfilling the responsibility to the employer and the responsibility to the client, human service professionals advise both of the conflict and work conjointly with all involved to manage conflict.

The Human Service Professional's Responsibility to Self

Statement 35. Human service professionals strive to personify those characteristics typically associated with the profession (e.g. accountability, respect for others, genuineness, empathy, pragmatism).

Statement 36. Human service professionals foster self-awareness and personal growth in themselves. They recognize that when professionals are aware of their own values, attitudes, cultural background, and personal needs, the process of helping others is less likely to be negatively impacted by those factors.

Statement 37. Human service professionals recognize a commitment to lifelong learning and continually upgrade knowledge and skills to serve the populations better.

References

Southern Regional Education Board. (1967). *Roles and functions for mental health workers: A report of a symposium.* Atlanta, GA: Community Mental Health Worker Project. Used with permission.

Index